REVISE EDEXCEL GCSE (9–1)

Geography B

REVISION GUIDE

Series Consultant: Harry Smith

Author: Rob Bircher

A note from the publisher

In order to ensure that this resource offers high-quality support for the associated Pearson qualification, it has been through a review process by the awarding body. This process confirms that this resource fully covers the teaching and learning content of the specification or part of a specification at which it is aimed. It also confirms that it demonstrates an appropriate balance between the development of subject skills, knowledge and understanding, in addition to preparation for assessment.

Endorsement does not cover any guidance on assessment activities or processes (e.g. practice questions or advice on how to answer assessment questions), included in the resource nor does it prescribe any particular approach to the teaching or delivery of a related course.

While the publishers have made every attempt to ensure that advice on the qualification and its assessment is accurate, the official specification and associated assessment guidance materials are the only authoritative source of information and should always be referred to for definitive guidance.

Pearson examiners have not contributed to any sections in this resource relevant to examination papers for which they have responsibility.

Examiners will not use endorsed resources as a source of material for any assessment set by Pearson.

Endorsement of a resource does not mean that the resource is required to achieve this Pearson qualification, nor does it mean that it is the only suitable material available to support the qualification, and any resource lists produced by the awarding body shall include this and other appropriate resources.

Contents

COMPONENT 1: GLOBAL GEOGRAPHICAL ISSUES

Hazardous Earth

1 Global circulation
2 Natural climate change
3 Humans and climate change
4 Tropical cyclones
5 Tropical cyclone intensity
6 Tropical cyclone hazards and impacts
7 Dealing with tropical cyclones
8 Tropical cyclones
9 Tectonics
10 Plate boundaries and hotspots
11 Tectonic hazards
12 Impacts of earthquakes
13 Impacts of volcanoes
14 Managing earthquake hazards
15 Managing volcano hazards

Development dynamics

16 What is development?
17 Development differences
18 Theories of development
19 Types of development
20 Approaches to development
21 Location and context
22 Globalisation and change
23 Economic development
24 International relationships
25 Costs and benefits

Challenges of an urbanising world

26 Urbanisation trends
27 Megacities
28 Urbanisation processes
29 Differing urban economies
30 Changing cities
31 Location and structure
32 Megacity growth
33 Megacity challenges
34 Megacity living
35 Megacity management

Extended writing questions

36 Paper 1

COMPONENT 2: UK GEOGRAPHICAL ISSUES

The UK's evolving physical landscape

37 Uplands and lowlands
38 Main UK rock types
39 Physical processes
40 Human activity

Coastal change and conflict

41 Geology of coasts
42 Landscapes of erosion
43 Waves and climate
44 Sub-aerial processes
45 Transportation and deposition
46 Landscapes of deposition
47 Human impact on coasts
48 Holderness coast
49 Coastal flooding
50 Coastal management

Fieldwork: coasts

51 Investigating coasts: developing enquiry questions
52 Investigating coasts: techniques and methods
53 Investigating coasts: working with data

River processes and pressures

54 River systems
55 Erosion, transportation and deposition
56 Upper course features
57 Lower course features 1
58 Lower course features 2
59 Processes shaping rivers
60 Storm hydrographs
61 River flooding
62 Increasing flood risk
63 Managing flood risk

Fieldwork: rivers

64 Investigating rivers: developing enquiry questions
65 Investigating rivers: techniques and methods
66 Investigating rivers: working with data

The UK's evolving human landscape

67 Urban and rural UK
68 The UK and migration
69 Economic changes
70 Globalisation and investment

Dynamic UK cities

71 A UK city in context
72 Urban change differences
73 City challenges and opportunities
74 Improving city life
75 The city and rural areas
76 Rural challenges and opportunities

Fieldwork: urban

77 Investigating dynamic urban areas: developing enquiry questions
78 Investigating dynamic urban areas: techniques and methods
79 Investigating dynamic urban areas: working with data

Fieldwork: rural

80 Investigating changing rural settlements: developing enquiry questions
81 Investigating changing rural settlements: techniques and methods
82 Investigating changing rural settlements: working with data

Extended writing questions

83 Paper 2

COMPONENT 3: MAKING GEOGRAPHICAL DECISIONS

People and the biosphere

84 Distribution of major biomes
85 Local factors
86 Biosphere resources
87 Biosphere services
88 Pressure on resources

Forests under threat

89 Tropical rainforest biome
90 Taiga forest biome
91 Productivity and biodiversity
92 Tropical rainforest deforestation
93 Threats to the taiga
94 Protecting the rainforest
95 Sustainable tropical rainforest management
96 Protecting the taiga

Consuming energy resources

97 Energy impacts
98 Access to energy
99 Global demand for oil
100 New developments
101 Energy efficiency and conservation
102 Alternative energy sources
103 Attitudes to energy

Extended writing questions

104 Paper 3 (i)
105 Paper 3 (ii)

Skills

106 Atlas and map skills
107 Types of map and scale
108 Using and interpreting images
109 Sketch maps and annotations
110 Physical and human patterns
111 Land use and settlement shapes
112 Human activity and OS maps
113 Map symbols and direction
114 Grid references and distances
115 Cross sections and relief
116 Graphical skills 1
117 Graphical skills 2
118 Numbers and statistics 1
119 Numbers and statistics 2

120 Answers

Edexcel publishes Sample Assessment Material and the Specification on its website. This is the official content and this book should be used in conjunction with it. The questions in Now try this have been written to help you practise every topic in the book. Remember: the real exam questions may not look like this.

Had a look ☐ Nearly there ☐ Nailed it! ☐

Global circulation

The atmosphere transfers heat around the Earth in a global circulation system. The way this circulation system creates areas of low pressure and high pressure explains why the Earth has areas with high rainfall (low pressure) and arid (high pressure) areas.

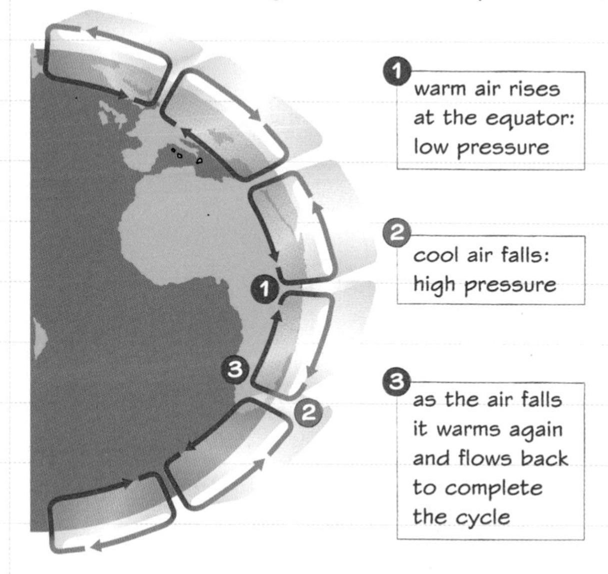

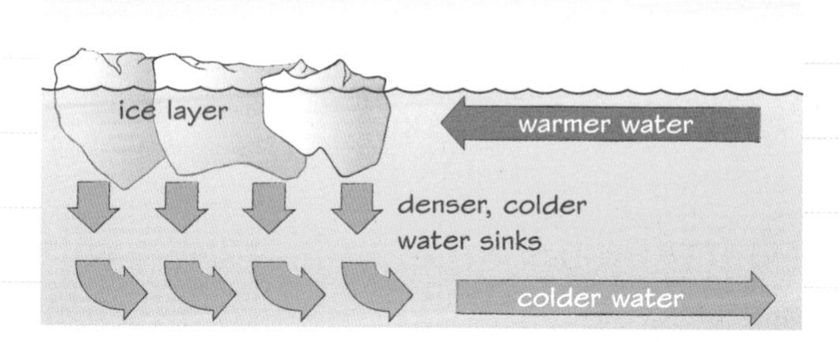

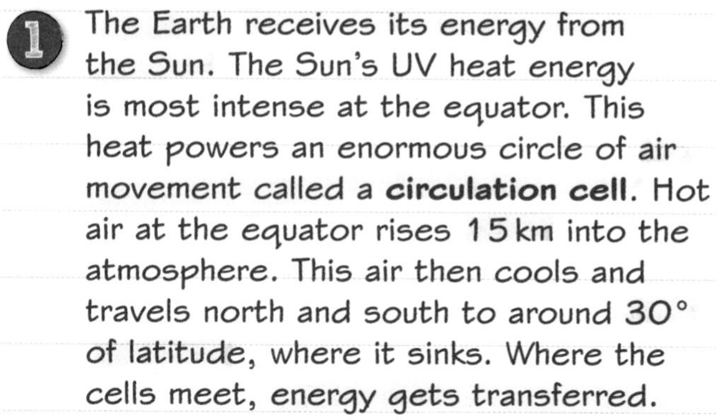

1 The Earth receives its energy from the Sun. The Sun's UV heat energy is most intense at the equator. This heat powers an enormous circle of air movement called a **circulation cell**. Hot air at the equator rises 15 km into the atmosphere. This air then cools and travels north and south to around 30° of latitude, where it sinks. Where the cells meet, energy gets transferred.

Ocean currents

Ocean currents also transfer heat around the globe. Some ocean currents are powered by wind resulting from the atmospheric circulation cells. Others are powered by density differences due to differences in water temperature and salinity. In the Arctic and Antarctic, the water gets very cold. This cold, salty, dense water sinks. As it sinks, warmer water from lower latitudes is pulled in. This is cooled too by the polar temperatures – and the cycle continues.

2 When warm air rises it creates low pressure. Rising air (becoming cooler and under lower pressure) cannot hold as much moisture and that's why precipitation is high at the equator. When the cool, dry air falls at 30° of latitude north and south, it creates high pressure. High pressure conditions have clear skies with little precipitation. These areas are often **arid**.

This climate graph is for Tindouf, a location in Algeria, North Africa, at a latitude of 27° north.

Worked example

Explain **one** reason why Tindouf has an arid climate. **(2 marks)**

Tindouf is in an area of high pressure created by an atmospheric circulation cell (the Hadley cell). Hot air pushed up at the equator loses its moisture and flows north, cooling and falling at around 30° north, creating high pressure and dry conditions.

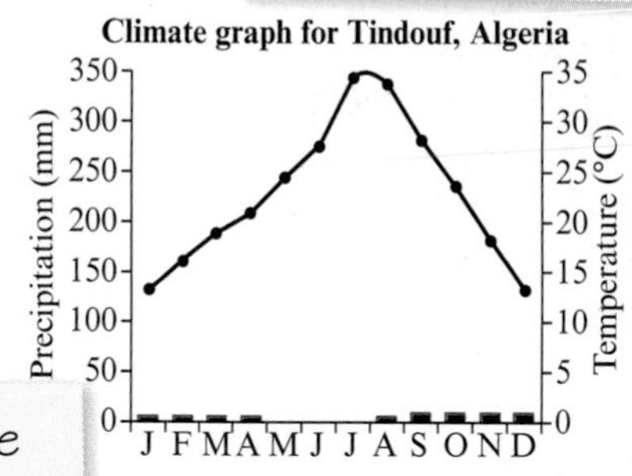

This answer makes good use of specialist terminology and precise details.

Now try this

Using the climate graph for Tindouf, calculate the annual temperature range. **(1 mark)**

Natural climate change

Although climate change is something humans have influenced, climate has changed for natural reasons many times in the Earth's history, on timescales from hundreds to millions of years.

This graph shows how the Earth's temperature has cooled and warmed over 450 000 years. It demonstrates **long-term** temperature changes due to natural causes.

The data for this graph comes from ice cores and fossils. **Ice cores** are cylinders of ice obtained by drilling through glaciers up to 3 km deep in ice up to 500 000 years old.

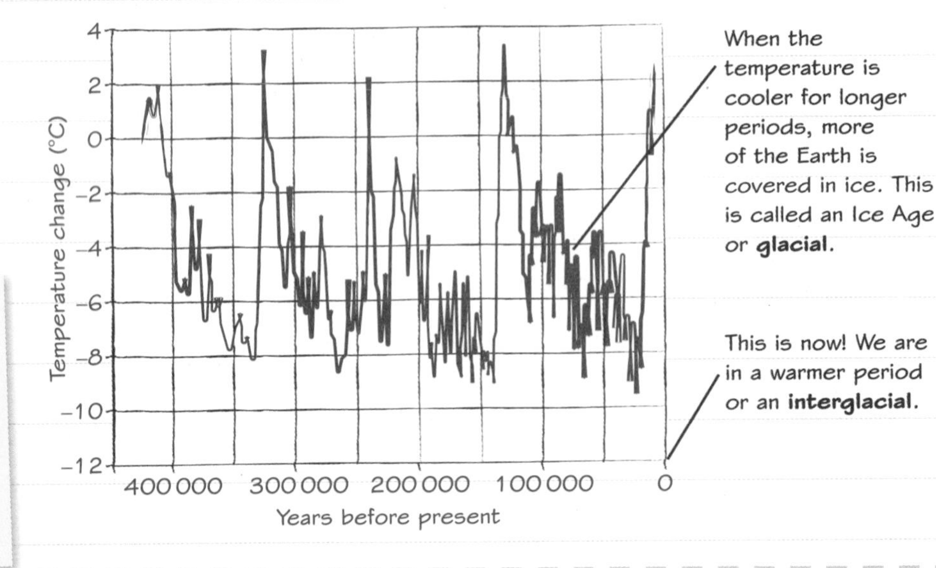

Some natural causes of climate change

- The Earth's orbit changes a small amount once every 100 000 years. These are known as **Milankovitch cycles**.
- The amount of energy radiated from the Sun changes over an 11-year cycle.
- Volcanic eruptions pump ash dust into the atmosphere causing a cooling effect.
- Large asteroid collisions can cause cooling as material blocks out the Sun. Asteroids hitting the Earth can cause huge fires which release massive amounts of CO_2 which subsequently has a warming effect.
- Ocean current changes can cause cooling and warming. In the UK, we have a warm and wet climate because of warm Atlantic currents. Sometimes the current shifts and we get a cooler climate for a short period of time.

Worked example

The graph opposite outlines UK climate changes since Roman times. Explain how **one** type of evidence is used to reconstruct how climate has changed in the past. **(3 marks)**

As snow falls it is converted to ice and isotopes of oxygen and hydrogen become trapped. Measuring the way their concentration has changed over time (using ice cores) provides a record of temperature change.

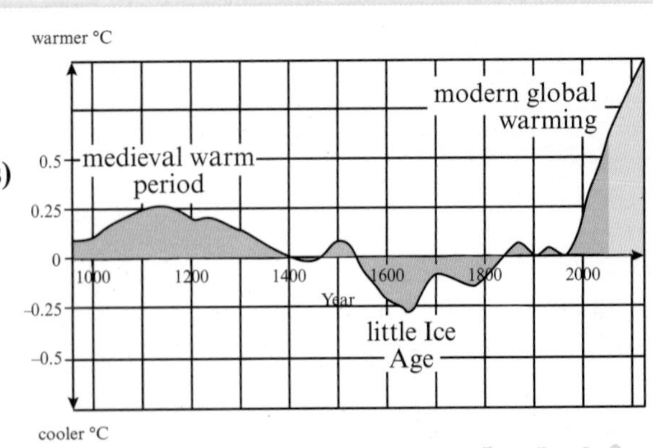

Now try this

The Quaternary glaciation (ice age) started 2.58 million years ago and has featured many glacial and interglacial events. When did the Quaternary glaciation finish? **(1 mark)**

☐ **A** 1 million years ago ☐ **B** 40 000 years ago ☐ **C** we are still in the Quaternary glaciation

Humans and climate change

Most scientists agree that the Earth's climate is warming and that the warming is in large part caused by human activity. You need to know how these human activities are causing warming, what the evidence is for this and why there are a range of projections for impacts on global temperature and sea-level rises.

Global warming is closely associated with rising atmospheric CO_2. This is a powerful greenhouse gas and is released by many human activities, including:

- industry
- transport
- energy production
- farming.

The enhanced greenhouse effect

- ✓ Heat (UV rays) from the Sun reaches the Earth's atmosphere; some is reflected back into space.
- ✓ The land and oceans absorb the heat.
- ✓ The land and oceans then radiate infrared heat back into the atmosphere.
- ✓ Greenhouses gases in the atmosphere trap some of the heat (necessary for life on Earth!).
- ✓ Human activity increases greenhouse gases in the atmosphere, leading to more warming.

What's the evidence? Six facts to remember

1. Global temperatures are rising – for example, in 2015 the average global temperature was 1°C above the average global temperature in 1850–1900.
2. Atmospheric CO_2 levels are rising in parallel with global temperatures. This is mostly due to human activity.
3. The oceans warmed by 0.11°C per decade between 1971 and 2010.
4. Sea levels rose globally by about 14 cm during the 20th century.
5. Arctic sea ice covers 13 per cent less of the sea each decade.
6. Extreme weather events have become more frequent: heat extremes are five times more common now than a century ago, for example.

Possible consequences of global warming

- biodiversity loss on land and in the oceans
- coastal flooding from sea-level rises
- more destruction from more frequent, stronger hurricanes
- more droughts, lasting longer
- spread of pests and diseases
- more flood from more frequent, heavier precipitation
- loss of glaciers would mean water supply problems in some areas
- changes in farming could affect food supplies

Worked example

Explain why projections for future global temperature rises and future sea-level rises may not be accurate. **(4 marks)**

The atmosphere and the oceans are highly complex systems. Computer models cannot always accurately predict how these systems will respond. For example, the oceans have absorbed more heat than was expected. Second, some natural events are hard to predict. For example, volcanoes can cause atmospheric cooling and the Sun can enter a cooler phase, affecting global temperatures.

Now try this

Study the diagram above. Choose **one** of the possible consequences of global warming shown. Explain why it is a possible consequence. **(2 marks)**

Had a look ☐ Nearly there ☐ Nailed it! ☐

Tropical cyclones

Tropical cyclones (violent tropical storms) are known by different names around the world – hurricanes, cyclones and typhoons. You need to know where and when they occur and about their key characteristics.

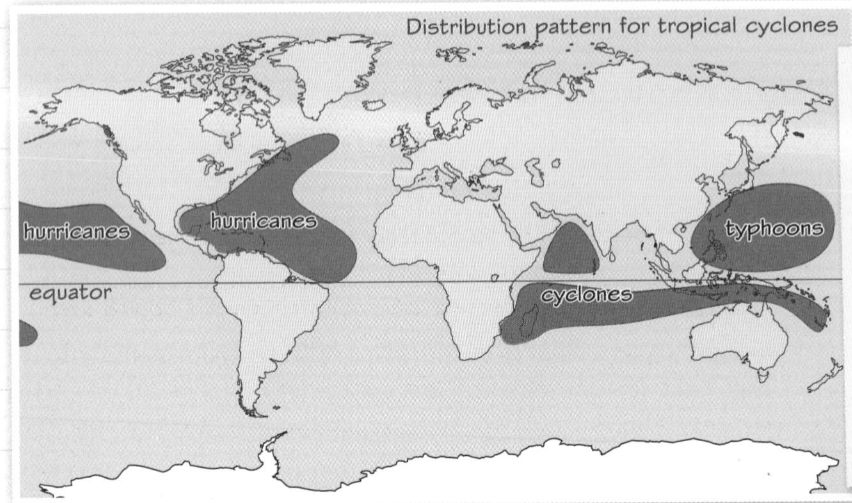

Tropical cyclones only form where seawater temperatures are above 26.5 °C. This limits:

- their geographical distribution (they occur in the tropics, starting between 5° and 30° of latitude)
- their seasonal distribution (they happen in summer and late autumn when seawater gets warmest).

Tropical cyclones have some key characteristics.

1. **Low pressure** – very warm, moist air rises through the atmosphere, sucking more air up behind it.
2. **Rotation** – the Earth's spin (Coriolis force) helps the rising air to spiral and drags in strong winds (there isn't much spin at the equator).
3. **Structure** – tropical cyclones form a cylinder of rising, spiralling air surrounding an eye of descending, high-pressure air. They are up to 640 km wide and 10 km high.

Tropical cyclone movements

- Tropical cyclones start in the tropics – warm water.
- They move westwards because winds blow from the east around the equator, and they spin away from the equator.
- Some tropical cyclones reach a belt of winds blowing from the west. This makes them change direction.
- Tropical cyclones can travel 640 km in a day.

Worked example

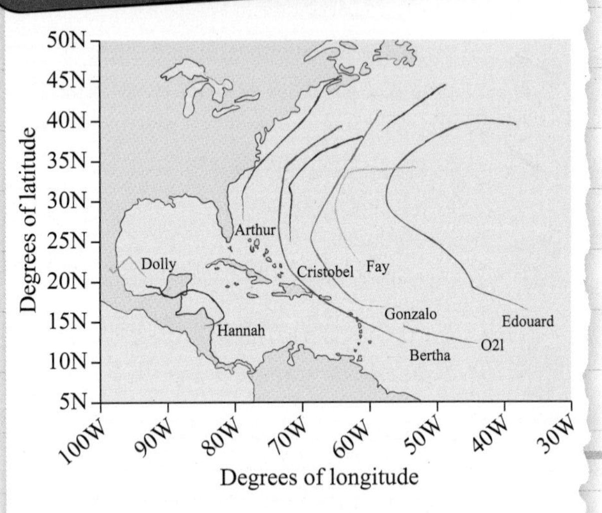

The chart above shows tropical cyclone tracks in the North Atlantic for July–September 2014. Using the chart, state the approximate latitude and longitude of the source of tropical cyclone Bertha. **(1 mark)**

12°N, 54°W

Remember that any exam questions can involve using your geographical skills as well as testing your geographical knowledge and understanding.

Now try this

The chart of tropical cyclones on this page is for July–September 2014. Explain **one** reason why tropical cyclones in the North Atlantic are most common in these months. **(2 marks)**

Tropical cyclone intensity

Some tropical cyclones are more intense than others, as categorised by the Saffir-Simpson scale. You need to know reasons why tropical cyclones **intensify** and **dissipate**.

The Saffir-Simpson scale

The Saffir-Simpson scale classifies tropical cyclones into five categories.

Category	Max. wind speed (km/hour)	Pressure (millibars)	Storm surge (metres)	Damage
1	119–153	980 and over	1.0–1.7	Some damage – trees lose branches, power lines come down
2	154–177	979–965	1.8–2.6	Roofs and windows damaged, some trees blown over, coastal flooding
3	178–208	964–945	2.7–3.8	Structural damage to buildings. Flooding over 1 m up to 10 km inland
4	209–251	944–920	3.9–5.6	Major devastation – destroys buildings, floods up to 10 km inland
5	252 or higher	<920	>5.7	Catastrophic – destruction up to 5 m above sea level. Mass evacuation needed

What makes tropical cyclones intensify and dissipate?

Tropical cyclones **intensify** when:

- water temperatures are warm – over 26.5 °C
- there is low wind shear
- there is high humidity.

Tropical cyclones **dissipate** (get weaker) when they:

- reach land because they lose energy (they are powered by warm water)
- move into areas of colder water
- run into other weather systems where winds are blowing in different directions.

How do tropical cyclones form?

How tropical cyclones form is not understood completely but six factors are very important.

1. Warm sea temperatures.
2. High humidity – there needs to be a lot of moisture in the atmosphere.
3. Rapid cooling – rising air must condense quickly to generate the huge amounts of energy powering a tropical cyclone.
4. Low wind shear – if winds are blowing in different directions up through the atmosphere, the cyclone won't form.
5. Coriolis force to give the cyclone spin – this isn't usually strong enough within 5° latitude of the equator.
6. Pre-existing low-pressure disturbances – tropical cyclones usually form when smaller storms come together.

Worked example

Explain **two** reasons why tropical cyclones do not form near the UK. **(4 marks)**

Tropical cyclones require water temperatures of over 26.5 °C to form, which is a lot warmer than UK sea temperatures (maximum around 17 °C). Tropical cyclones need wind speeds and directions to be similar up through the atmosphere. The UK's climate is dominated by weather fronts where wind speeds and directions change through the atmosphere.

Now try this

The maximum recorded wind speed for Typhoon Haiyan (November 2013) was 315 km/h, with average speeds of 280 km/h. State which category this was on the Saffir-Simpson scale. **(1 mark)**

Tropical cyclone hazards and impacts

Tropical cyclones can be major natural hazards that can have devastating impacts on people and places. You need to know about the physical hazards and impacts.

Some tropical cyclone statistics

- Hurricane Katrina (USA, 2005) caused $108 billion of damage, with 1836 deaths from a **storm surge** up to 8.5 m high.
- The highest recorded number of deaths caused by a tropical cyclone to date is 500 000 people in Bangladesh, 1970.
- In 2015, Cyclone Pam in the South Pacific was the most intense tropical cyclone measured in the southern hemisphere, with a pressure of 896 mb.

Physical hazards

Tropical cyclones are very destructive with strong winds (100 km to 300 km per hour) and intense rainfall (at least 100 mm in 24 hours).

However, the most dangerous physical hazards for humans are **secondary hazards** triggered by the tropical cyclone:

- ✓ storm surges
- ✓ landslides
- ✓ coastal flooding.

Storm surges

- As the tropical cyclone moves toward the coast, the sea gets shallower. Water pushed up by the wind in front of the storm has nowhere to go but up and onto the land. Storm surges have reached 12 m high!
- The low atmospheric pressure in the tropical cyclone also increases the surge.
- Large, slow-moving tropical cyclones cause the biggest surges.

The storm surge caused by Typhoon Haiyan in November 2013 reached heights of over 6 m.

Coastal flooding

Storm surges can flood large areas of the coast if the land is low-lying. Coastal flooding has reached 20 km inland in some cases.

Landslides

Rainfall is heavy and intense during tropical cyclones. 200–300 mm of rain falling in just a few hours causes flash flooding and triggers landslides on unstable slopes (this instability is often linked to deforestation).

Worked example

Explain the impact that tropical cyclones can have on people and the environment. **(4 marks)**

Category 4 and 5 tropical cyclones have wind speeds that can knock down houses, leaving people homeless, as well as destroying power lines, leaving people without power, and blocking roads with fallen trees and debris.

However, it is storm surges caused by tropical cyclones that do the most damage to people and the environment. Storm surges drown people, sweep away buildings, destroy crops, contaminate agricultural land with salt, and cause sewage spills that threaten public health through waterborne diseases, such as typhoid.

Now try this

Cyclone Pam reached a pressure of 896 mb. State which category on the Saffir-Simpson scale it would match for atmospheric pressure (see page 5 for the Saffir-Simpson scale). **(1 mark)**

Dealing with tropical cyclones

You need to know why some countries are more vulnerable than others to tropical cyclones, and the different ways that countries can prepare for and respond to them.

These satellite images show two sections of a coastal area in the Philippines, one before and one after Typhoon Haiyan (2013)

Low-lying coasts are vulnerable to tropical storms.

Debris blocks roads making it hard to rescue survivors. Developing countries may have to wait for international assistance to clear debris.

Although high winds (315 km/h) caused physical hazards, the main destruction was from a 7-metre storm surge. This coastline had no defences against this.

Weather forecasting and satellites

Satellites are used to spot a tropical cyclone forming and track its progress. Forecasters can predict its track and estimate likely storm surge heights and rainfall levels.

WHAT SHOULD I DO?

ARE YOU UNDER THREAT?

Find out from your local Emergency Services whether you are in a surge-prone area. Work out how to get to your nearest shelter safely.

ARE YOU READY TO EVACUATE?

Plan what you will do if you have to evacuate. Will you have essential medicines? Vital documents? What about your pets? Discuss your plans with your local council.

TIME TO EVACUATE!

Be prepared to evacuate as soon as you are advised to. This makes it easier for Emergency Services to manage the difficult task of moving a lot of people all at once, especially if the weather is getting worse. When a cyclone threat develops, keep listening to official warnings of high tides and coastal flooding from the Bureau of Meteorology.

Different factors combine to make countries more or less vulnerable:

- **Physical** – low-lying coastal areas (where lots of people live), e.g. low-lying Pacific islands.
- **Social** – poor areas are often hit worst because poor people live at high density, in low-lying areas, in poor quality housing.
- **Economic** – rich, developed countries have better prediction, protection and evacuation technology. This makes them less vulnerable.

This is an example of government evacuation instructions, from Australia. They include:

- prediction services and warnings from the Bureau of Meteorology
- evacuation infrastructure: high-ground shelters for people in surge-prone areas
- information on preparing for evacuation.

Worked example

Satellite images are useful for identifying tropical cyclones as they form and monitoring their movements. Which one of the following is the correct term for this movement? **(1 mark)**

☐ path ☒ track ☐ route ☐ area

Now try this

Calculate the death toll of Hurricane Katrina (1836) as a percentage of the death toll of Cyclone Nargis (138 866). **(1 mark)**

Tropical cyclones

Located example

You need **located examples** of the effectiveness of preparations for, and responses to, tropical cyclones in **one** developed country and in **one** developing/emerging country. Revise the case studies you did in class!

Developed country: USA

In August 2005, Hurricane Katrina caused 1836 deaths in Louisiana, USA: mostly in the city of New Orleans. This was shocking!

Katrina was only a category 3 when it hit New Orleans.

The deaths were mainly due to failures in storm surge defences. New Orleans' levee system was old and hadn't been sufficiently maintained. Pumping stations were flooded and failed to work.

The evacuation procedure had problems. Highways out of New Orleans jammed. Public transportation was not used. Shelters did not have enough food.

Improvements since Hurricane Katrina

All the city's 400 km of levees have been made much higher and much stronger

All the city's 78 floodwater pumping stations have been made floodproof.

The Lake Borgne surge barrier has been built to protect New Orleans – the largest storm-surge barrier in the world.

Upgrading these responses cost **$14 billion**.

New funding has been spent on search and rescue teams and city residents now get evacuation updates by text message.

Emerging country: Myanmar

In May 2008, Cyclone Nargis caused 138 866 deaths in Myanmar (Burma).

Cyclone Nargis was a category 5 tropical cyclone. It had been tracked by weather forecasters in India and Bangladesh, who informed the Myanese/Burmese government.

There were no evacuation procedures. There were no defences against storm surges. Houses were made of very weak materials. One million people were homeless after a 7.6 m storm surge.

The government had limited ways of helping the 2.5 million people affected by the cyclone but still refused to let foreign aid workers in to help for a whole week.

Improvements since Cyclone Nargis

International aid organisations have built much stronger schools, hospitals and homes.

New sanitation systems have been installed by international aid organisations that are raised up above likely flood levels.

Local communities now have special committees to organise evacuations.

The government has built 20 cyclone shelters, each for 500 people.

Two people in each community are given a mobile phone so they can relay cyclone warnings.

There are plans to regrow the mangrove forests that used to protect the coast.

Worked example

Describe **one** way a named developing country prepared for a tropical cyclone. **(2 marks)**

The government of the Philipines built strong brick typhoon shelters on higher ground. However, when Typhoon Haiyan hit Tacloban, the storm surge was higher than the shelters and too strong for them to resist.

Now try this

'Developed countries provide much better protection against tropical cyclones than developing or emerging countries.' Assess this statement with reference to named examples. **(8 marks)**

Tectonics

Why does the Earth have plates? How do they move? The answers can be found in the structure of the Earth and the composition of its different layers.

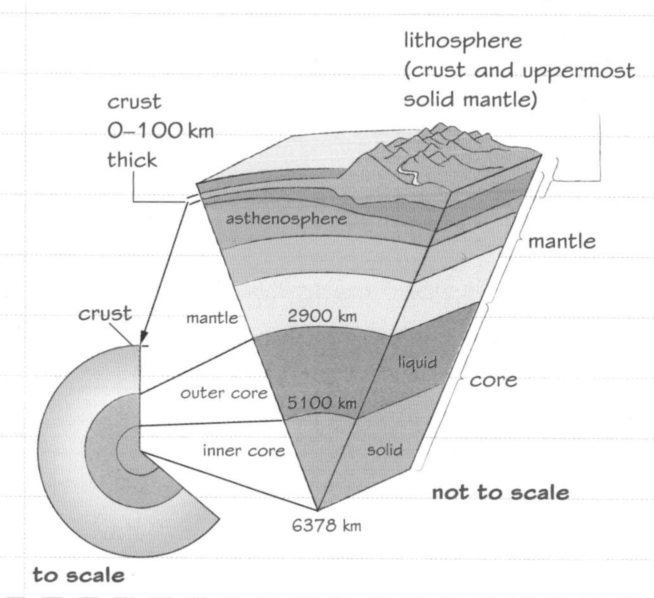

The Earth is made up of a series of layers, like the layers of an onion. There are three main layers: crust, mantle and core.

- The **crust** is solid and rigid – tectonic plates.
- The tectonic plates move on top of the **asthenosphere** – a solid but 'plastic' layer under such high pressure that the rock flows.
- The lower layer of the **mantle** is liquid magma at 3000°C.
- The outer **core** is liquid iron and nickel. Temperatures are 4000–6000°C.
- The inner **core** is iron at temperatures of 5000–6000°C. The pressure is so high that this iron is solid.

How convection currents contribute to plate movement

1. The core heats the molten rock in the mantle to create a **convection current**.
2. Heated rock from the mantle rises to the Earth's surface.

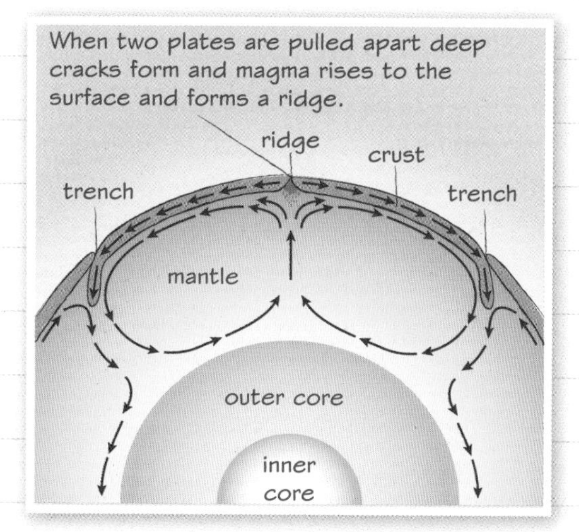

3. At the surface the convection current moves the tectonic plates in the crust.
4. Molten rock cools and flows back to the core to be reheated.

Worked example

Which of the following is the main heat source that powers convection? **(1 mark)**

☐ magma ☒ radioactive decay
☐ continental drift ☐ subduction

Multiple choice questions often include incorrect answers that look like they could be correct, so check carefully.

Density

There are two main types of crust: continental crust and oceanic crust.

- Continental crust is mainly granite, while oceanic crust is mainly basalt.
- Continental crust is **less dense** than oceanic crust.
- Both continental crust and oceanic crust are **less dense** than the rocks of the asthenosphere.

Now try this

Explain why radioactive heat is important to tectonics. **(2 marks)**

Plate boundaries and hotspots

There are three types of plate boundary. They are linked to where volcanoes and earthquakes occur. Hotspots are also places of high volcanic activity.

The three main types of **plate boundaries** are:

1 Convergent plate boundaries

- **Example:** Nazca Plate and South American Plate
- Two plates collide, one plate flows beneath another (**subduction**)
- Many earthquakes and volcanoes

Collision plate boundaries

- ✓ **Example:** Indo-Australian and Eurasian plates
- ✓ Two continental plates collide and the two plates buckle
- ✓ Many earthquakes

2 Divergent plate boundaries

- **Example:** Eurasian and North American plates
- Rising **convection currents** pull crust apart forming volcanic ridge, e.g. Mid-Atlantic Ridge

3 Conservative plate boundaries

- **Example:** San Andreas Fault, California
- Two plates slide past each other
- Earthquakes

Main tectonic plates

This is a diagram of the global tectonic plates. It shows the three main types of plate boundaries and the direction of the plate movement, as well as the collision plate boundaries. Plate movements are complex and not yet fully understood.

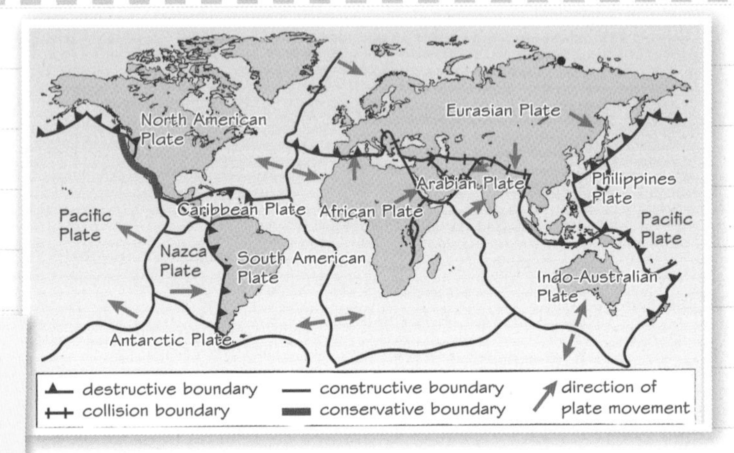

Hotspots are points on the Earth's crust with very high heat flow, which is linked to increased volcanic activity. Some are on plate boundaries but, strangely, many are not.

Worked example

Look at the map of tectonic plates.

Which of the following is the volcano Kilauea formed on? **(1 mark)**

☐ **A** convergent plate boundary
☒ **B** hotspot
☐ **C** conservative plate boundary
☐ **D** divergent plate boundary

Make sure that you look at the resource indicated by the question, and locate the correct volcano.

Now try this

Remember that **annotations** are **explanatory** notes, not just labels.

The island of Lanzarote off the Atlantic coast of North Africa has formed over a hotspot.
Explain how volcanoes form over hotspots. Use an annotated diagram or diagrams in your answer. **(4 marks)**

Tectonic hazards

Volcanic and earthquake hazards vary and cause different types of damage and devastation, depending on a range of factors.

Shield volcanoes:

- are found on constructive plate boundaries or hotspots
- are formed by eruptions of thin, runny lava which flows a long way before it solidifies
- have gently sloping sides and a wide base
- contain basaltic magma which is very hot with low **silica** and gas content
- erupt frequently but not violently.

Composite volcanoes:

- are found on destructive plate boundaries
- are formed by eruptions of viscous, sticky lava and ash that don't flow far
- have steep sloping sides and a narrow base
- are made up of layers of thick lava and ash
- contain andesitic magma which is less hot than basaltic magma but which contains lots of **silica** and gas
- erupt infrequently but violently, including pyroclastic flows (mix of ash, gases and rock).

Energy from earthquakes

The energy produced by earthquakes is at its greatest at two key points:

1. **Epicentre** – the point on the surface directly **above** the **focus**
2. **Focus** – the central point of the earthquake deep under the surface, where the earthquake actually happens!

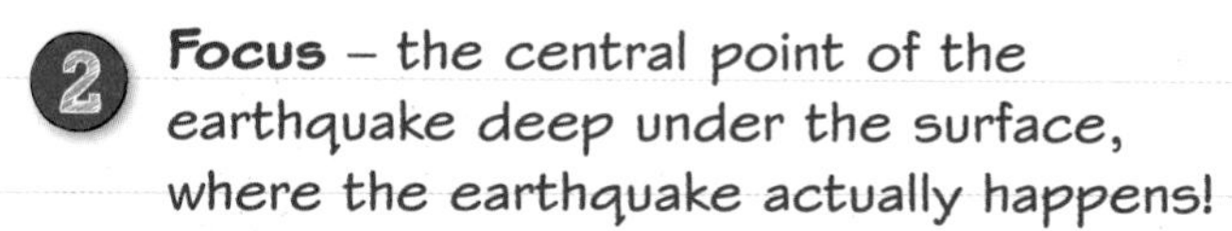

earthquake magnitude:
a measure of the size of the earthquake

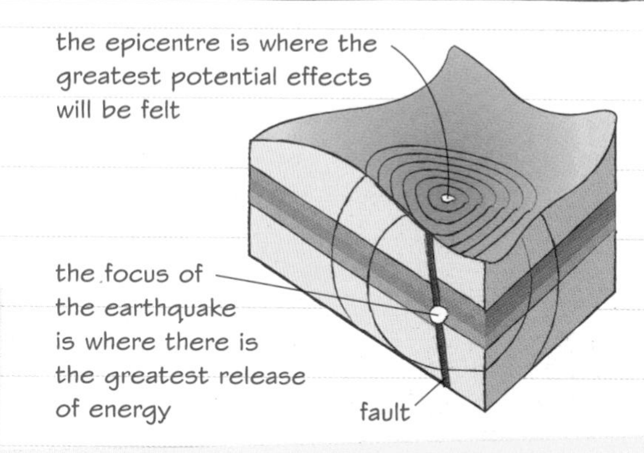

Earthquakes occur on all three plate boundary types but are most common on convergent boundaries. Severe earthquakes are high on the Richter scale, have a shallow focus (near the surface) and occur in sands and clays, which vibrate more than solid rock.

This answer is very full and the student's comparison of shield and composite volcanoes has made the points in the answer easier to make and clearer to understand.

Worked example

Explain **one** reason why composite volcanoes are often more dangerous to humans than shield volcanoes. **(3 marks)**

While shield volcanoes contain basaltic magma, which flows smoothly and is not full of gas, composite volcanic eruptions involve andesitic magma, which contains a lot of silica. This clogs up the volcanic vent so pressure builds up in the volcano, along with large amounts of gas that make it explosive, and these lead to very violent eruptions.

Now try this

Define the term 'tsunami'. **(2 marks)**

Give **two** examples of recent tsunamis. **(2 marks)**

Explain how a tsunami is caused. Use a diagram in your answer. **(4 marks)**

Had a look ☐ Nearly there ☐ Nailed it! ☐

Impacts of earthquakes

You need to know a located example from a developed country and from a developing or emerging country for this topic.

Primary and secondary impacts

The way that earthquakes affect people and property in developed countries can be different from the effects in developing or emerging countries.

Don't revise this page if you studied volcanoes for your located examples for this topic. Check your notes.

Primary impacts

Primary impacts are things that happen immediately as a result of an earthquake, for example:

- deaths and injuries
- destruction of buildings or damage to buildings
- destruction or damage to roads, railways, bridges.

How bad primary effects are depends on a mix of physical and human factors, e.g. how strong the earthquake is, and whether it happens in a crowded city or a sparsely populated rural area.

Secondary impacts

Secondary impacts are the after-effects of earthquakes, for example:

- fires caused by fractured gas pipes and broken electricity pylons
- landslides on steep or weak slopes
- spread of disease when sanitation breaks down
- tsunamis, when the earthquake occurs offshore.

Secondary effects may have a bigger impact in poorer countries because they do not have the money to prepare people for earthquakes or protect buildings and infrastructure.

Worked example

Explain the secondary impacts of a named earthquake event that you have studied. **(4 marks)**

15 900 people were killed by the Tohoku earthquake in Japan in 2011 (magnitude 9.0). Almost all of them were killed by a secondary effect: a tsunami triggered by the earthquake. Millions of residents lost water supplies and power supplies. The tsunami also caused damage to the Fukushima nuclear power station, causing nuclear meltdowns and radiation leaks. 300 000 people had to be evacuated from Fukushima district.

The earthquake was caused by the subduction of the Pacific Ocean plate under the plate that carries Japan. The earthquake had a shallow focus – 30 km below the surface – which lifted a large section of seafloor by 10 m. This displaced the water column above, causing the tsunami.

A tsunami is a secondary impact of an earthquake, so this is a good example to use.

Now try this

Answer the question in the worked example for yourself using the located example and information you covered in class (e.g. you may have studied the 2004 Indian Ocean earthquake and tsunami). If you did the Japan earthquake too, answer the worked example question for your developing or emerging country located example.

Impacts of volcanoes

You need to know a located example from a developed country and from a developing or emerging country for this topic.

Primary and secondary impacts

The way that volcanoes affect people and property in developed countries can be different from the effects in developing or emerging countries.

Don't revise this page if you studied earthquakes for your located examples for this topic. Check your notes.

Primary impacts

Primary impacts are things that happen immediately as a result of a volcanic eruption, for example:

- deaths and injuries
- destruction of buildings or damage to buildings
- destruction or damage to roads, railways, bridges, farmland.

How bad primary impacts are, depends on a mix of physical and human factors – for example, the type and intensity of the eruption, and whether it happens in a crowded city or a sparsely populated rural area.

Secondary impacts

Secondary impacts are the after-effects of volcanic eruptions, for example:

- atmospheric pollution caused by ash
- mudflows (lahars)
- landslides of volcanic debris
- flooding if lava flows block rivers
- tsunamis from landslides or collapses caused by the eruption.

Impacts of volcanic eruptions may be bigger in some developing countries and emerging countries than in developed countries because developed countries are generally able to afford better monitoring, prediction, warning, and preparation and response systems.

Worked example

Explain the primary impacts of a named volcanic event in a developing country you have studied. **(4 marks)**

The volcanic eruption in Montserrat in 1995 had many primary impacts, including the destruction of Montserrat's capital city, Plymouth, as it was covered in ash; the destruction of most of the island's crops by ash; and the destruction of forests by fires ignited by gases from the volcano.

Montserrat's volcano is a composite volcano containing andesite lava, which is very thick and blocks volcanic vents, leading to pressure build-up and explosive eruptions. A dome of lava built up on the volcano, which suddenly collapsed, with the explosion producing vast amounts of ash and debris, which then fell back to the surface, forming thick deposits.

The details of this answer are all relevant to the question, which is the right approach.

Now try this

Answer the question in the worked example for yourself, using the located example and information you covered in class. If you did the Montserrat volcano too, answer the worked example question for your developed country located example.

Managing earthquake hazards

Located example You need to know a located example about managing earthquake hazards from a developed country and from a developing or emerging country for this topic.

Don't revise this page if you studied volcanoes for your located examples for this topic. Check your notes.

The way earthquakes are managed in developed countries can be different from how they are managed in developing or emerging countries.

What is needed after an earthquake?

- Trained volunteers to help the injured people and to clear away the debris.
- Clean water to prevent the spread of disease.
- Food because often shops, towns, roads and farms have been damaged.
- Radio communication because phones will often not work.
- Medical help to care for the injured people.
- A plan to evacuate the area if needed.

Buildings in earthquake areas

How to strengthen a building.

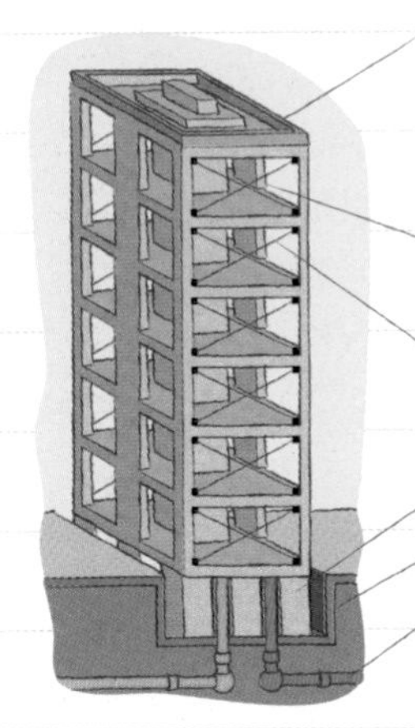

Worked example

Describe **two** differences in the ways in which buildings are made earthquake-resistant in developed countries and in emerging or developing countries. **(2 marks)**

In developed countries there is more money available to spend on making buildings earthquake-resistant, for example by using rubber and steel foundations which can move slightly. Another difference is that buildings in the developing world are more likely to be made from timber. A cheap way to make these buildings more earthquake-resistant is to add diagonal braces that reduce movement.

As well as describing two ways buildings are made earthquake-resistant, the question also wanted differences between developed and developing/emerging countries.

Remember to read questions carefully. If you are asked for short-term relief efforts, then do not write about long-term planning.

Now try this

Compare the short-term relief efforts after an earthquake in a developed country that you have studied and in a developing or emerging country that you have studied. **(5 marks)**

Managing volcano hazards

Located example

You need to know a located example from a developed country and from a developing or emerging country for this topic.

Don't revise this page if you studied earthquakes for your located examples for this topic. Check your notes.

The way volcano hazards are managed in developed countries can be different from how they are managed in developing or emerging countries.

Predicting eruptions

There are ways to monitor changes in volcanoes to help predict an eruption. Two of these are:

- tiltmeters that check for bulges on volcano slopes
- satellites that monitor changes in heat activity.

Monitoring is expensive and poorer countries often cannot afford it.

Preparing for eruptions

There are ways to prepare for volcanic eruptions to reduce their impact on people and property.

Three of these are:

- setting up an exclusion zone
- evacuating people from inside the exclusion zone
- organising barriers to divert lava flows or water-spraying equipment to halt flows.

Worked example

The image opposite is of a **Global Positioning System (GPS)** station located on a plate boundary in the USA.

Explain why prediction of volcanic eruptions is often better in developed countries than in some developing or emerging countries **(4 marks)**.

GPS prediction technology is expensive to install in the first place, but what developing countries often have most problems with is the cost of maintaining the equipment over time.

Governments run volcano prediction projects, but in developing countries governments cannot afford to pay high wages. The engineers and technical analysts needed for monitoring and analysis may go to work in better-paid jobs in the private sector.

This answer makes two supported points, which helps with the explanation.

Now try this

Describe **one** preparation that could be made by people living close to a dormant volcano in case of eruption.

(2 marks)

What is development?

Development is a term that measures how advanced a country is compared to others. It is about the standard of living in a country – whether people can afford the things they need to survive. However, it's not just about money. Development also includes the quality of life within a country.

Measuring development

The level of development in a country or region can be measured using statistics for **economic indicators** and **social indicators.** Some things such as birth rate are easy to measure but others, such as how safe people feel, are more difficult to quantify but some countries try to do this too. Bhutan measures its Gross National Happiness!

Some factors to consider when evaluating development

Economic	Physical well-being	Mental well-being	Social
• income • type of industries • security of jobs	• diet • access to clean water • environment (including climate, hazards, etc.)	• freedom • security • happiness	• access to education • access to health care • access to leisure facilities

Economic indicators

- **GDP** – **Gross Domestic Product** is the total value of goods and services produced by a country in a year. It's often divided by the population of that country to give GDP per capita.
- **HDI** – the **Human Developement Index** puts together measurements of a country's gross national income per capita (like GDP), life expectancy and years in education to provide a figure that represents the country's development level.

These have limitations because:
- all measures of development show averages only
- data do not show everything and are not always accurate. For example, GDP doesn't include the cash economy.

Political indicators

Political indicators show what a government is likely to be doing for its country.

- Is it well governed?
- Is there free speech?
- Is there corruption?

Worked example

Describe **one** example of an economic measure of development. **(2 marks)**

GDP is an economic measure of development. It is the total value of goods and services a country produces in a year.

Now try this

1 Explain some of the problems of **only** using economic measures of development. **(4 marks)**

2 Explain why GDP **per capita** is a better indicator of development than just GDP. **(2 marks)**

Development differences

Countries at different levels of development have differences in their demographic data. Knowing how to interpret population pyramids is important for this topic.

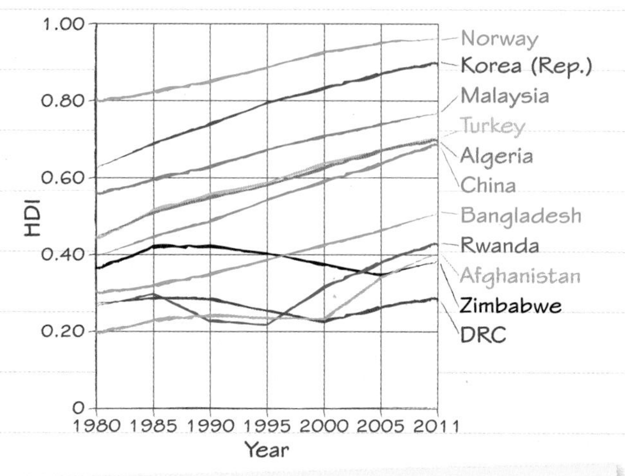

The graph shows that human development is not equal in all countries.
There are many different reasons for inequality.

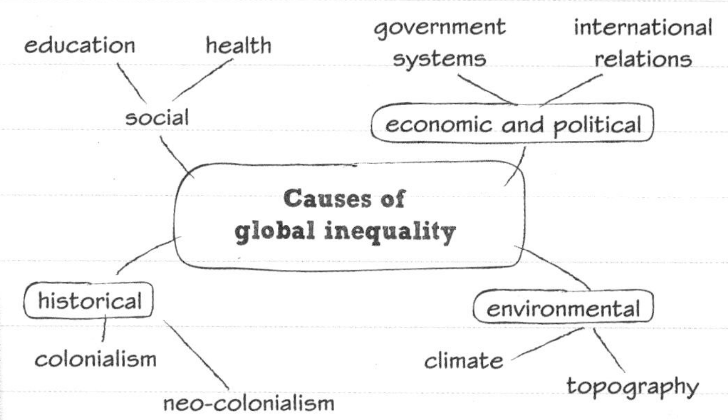

HDI scores each country between 0 and 1.
- 0.80 and over = high development
- 0.50–<0.80 = medium development
- <0.50 = low development

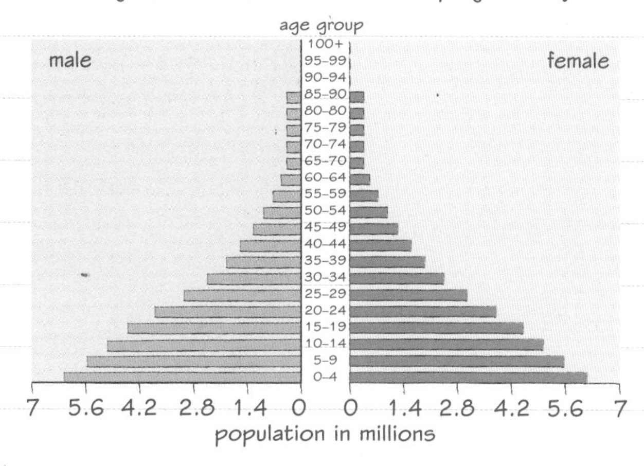

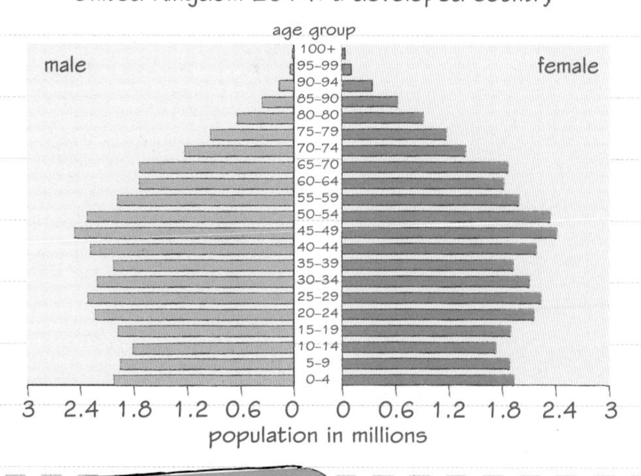

Population pyramids

- Emerging countries have high **fertility rates** – which gives their pyramids a broad base.
- Developed countries have lower fertility rates – a narrower base than sides. Emerging countries have a **youthful** population: most people are under 30.
- Developed countries have an **ageing** population: fewer young people, and increasing **life expectancy**.

Worked example

Explain why infant mortality rates vary between countries. **(4 marks)**

If there is not enough money to pay for medical care, infant mortality will be higher than where healthcare is well funded. If new mothers have not been educated about baby hygiene, infant mortality will be higher than where they have.

Environmental reasons may also be involved. For example, in tropical climates diseases like malaria cause many infant deaths.

Now try this

Explain **two** ways in which population structure can influence social issues. **(4 marks)**

Theories of development

There are many different theories to explain why societies develop. One of these is Rostow's **modernisation theory** and another is Frank's **dependency theory**.

Rostow's modernisation theory

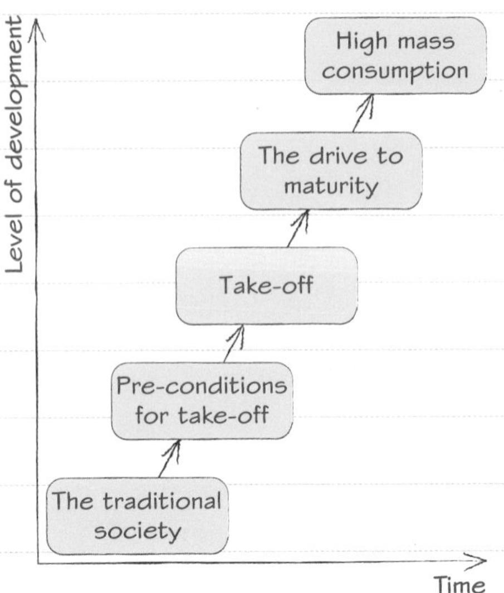

Problems

- It assumes that all countries start at the same level of development.
- It doesn't consider the quality or quantity of a country's resources, population or climate/natural hazards.
- It's out of date and based on the 18th- and 19th-century development of European countries.
- It fails to consider that European development came at the expense of other countries (colonisation).

Frank's dependency theory

This is the idea that developing countries can't develop because they are dependent on developed countries. The most developed countries have the economic and political power to exploit less developed countries and impose trade barriers and conditions for loans that hinder development.

Problems

- It was written in the 1950s so is outdated – today, some less developed countries are developing very quickly, e.g. China and India, which may show the dependency theory doesn't work (or only applies to some places).
- It doesn't take account of other factors which may limit development, such as natural disasters, lack of resources, conflict, etc.

Worked example

Study the diagram opposite, which represents how global income is divided between five equal shares of the world's population.

Explain one way that Frank's dependency theory could help explain this unequal distribution of global income. **(4 marks)**

The quintile of the richest 20 per cent owns over 80 per cent of global wealth according to this diagram, which dependency theory would explain as being due to the 'core' of richest countries exploiting their control of the world market system to keep poorer countries in the position of supplying raw materials which are then used to make the richest 20 per cent even richer.

Quintiles divide a total into five equal shares: a good term to use in this answer.

The distribution of global income in 2013

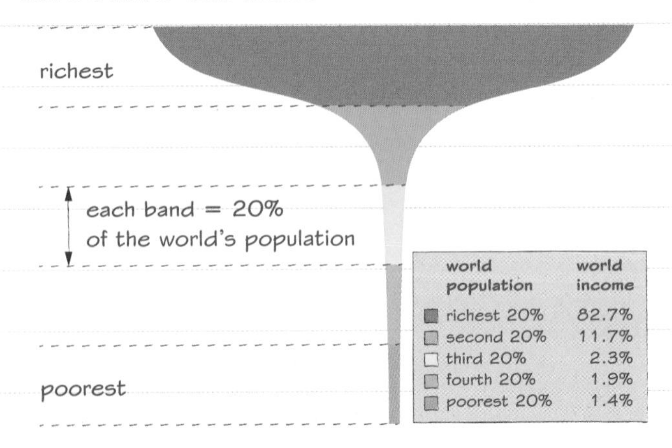

world population	world income
richest 20%	82.7%
second 20%	11.7%
third 20%	2.3%
fourth 20%	1.9%
poorest 20%	1.4%

Now try this

1 Identify two problems with Rostow's modernisation theory. **(2 marks)**

2 Explain why the dependency theory assumes that 'developing' countries will never become developed countries. **(4 marks)**

Types of development

You need to know about differences between top-down and bottom-up development, about **globalisation** and its differing impacts on development.

Top-down development

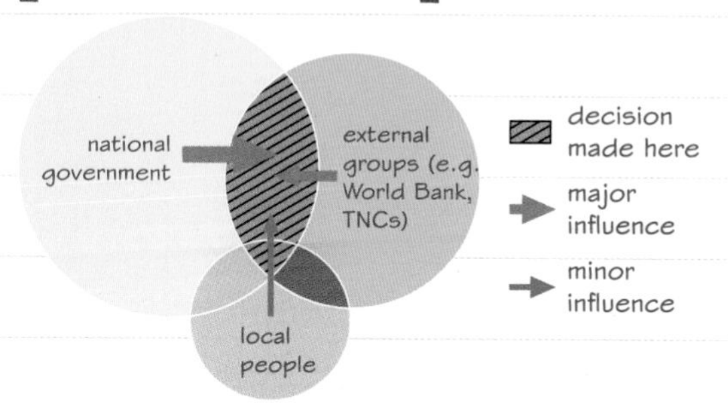

- Large-scale projects that aim at national-level or regional-level development.
- Very expensive projects often funded by international development banks.
- Sophisticated technology that needs experts to install and maintain.

Bottom-up development

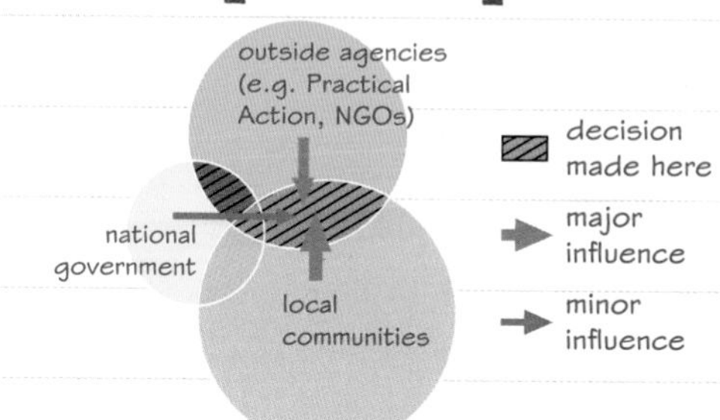

- Local-scale projects that aim to benefit a village or small group of communities.
- Very cheap compared to top-down, but usually funded by the village itself.
- Straightforward technology that local people can learn to operate and repair.

Globalisation – what is it?

Economic interests and a desire to make profits have encouraged companies from developed countries to produce in countries where labour is cheap and then to sell the products all over the world.

→ This pulls countries together in a global economy (globalisation).

→ Trade connections between countries has led to **interdependence** between them. Countries trade because one country has something the other country doesn't.

This word cloud shows exporting countries in 2014.

This answer identifies that processes boost global economic development while also increasing inequality between countries. Globalisation.

Worked example

Study the word cloud. The size of a country's name represents the amount they export.
Explain why some countries have a bigger share of global trade than others. **(4 marks)**

One reason is that most developing countries export raw materials. **Transnational corporations (TNCs)** buy raw materials cheaply from developing countries, and use them to make manufactured products in other countries. Manufactured products can be exported for more money.

China is an emerging country with a very big share of global exports because TNCs in developed countries have moved manufacturing there to benefit from cheaper wages.

Now try this

Explain **one** way that governments contribute to globalisation. **(2 marks)**

Approaches to development

You need to know advantages and disadvantages of **three** different approaches to development: NGO-led intermediate technology, IGO-funded large infrastructure, and investment by TNCs.

Definitions

NGO – Non-governmental organisation (e.g. the charity Action Aid)

IGO – Intergovernmental organisation (e.g. the United Nations)

intermediate technology – simple technology that local people can operate and maintain themselves

TNC – Transnational corporation

Solar cooking – an example of intermediate technology

NGO-led intermediate technology

- 👍 Targeted at specific needs of local people (e.g. supply of clean water from a new well)
- 👍 Generates jobs among local people (e.g. repairing technology, training users)
- 👎 Governments often rely on NGOs instead of developing their own systems to help their people
- 👎 Lack of data about how successful NGO schemes really are; not as **accountable** as IGOs

IGO-funded large infrastructure

- 👍 Can access very large sums of money from IGOs like the World Bank
- 👍 Large infrastructure developments can benefit hundreds of thousands of people
- 👎 High-tech solutions can be costly to maintain, and may fail if funds run out
- 👎 Many local people may not actually benefit (e.g. if they have to move because a major new dam project will flood their area)

Worked example

Study the map opposite, which shows investment in selected African countries from US TNCs.

Explain **one** reason why US TNCs might want to invest in African countries. **(3 marks)**

Countries like Nigeria and South Africa have valuable resources – for example, oil and diamonds. TNCs based in the USA can invest in these countries in order to get access to these resources.

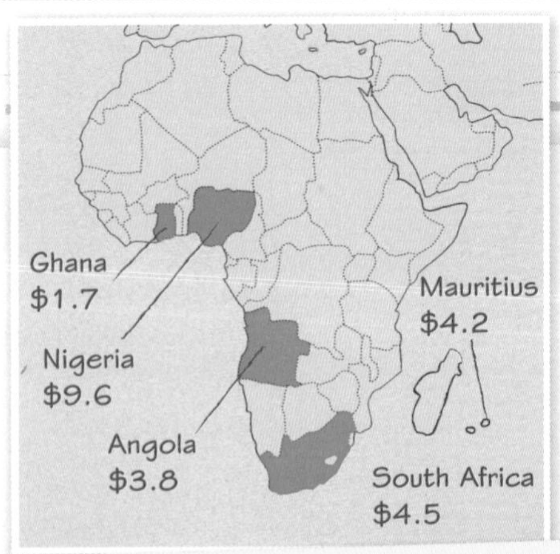

US TNC investment in Africa (in $ billions)

Now try this

Explain the disadvantages of TNC investment in developing or emerging countries. **(4 marks)**

Location and context

You will have studied one emerging country in terms of how it is managing to develop. There are some key facts you need to know about your emerging country.

You need to know about where your country is and reasons why it is **significant** in its region and in the world.

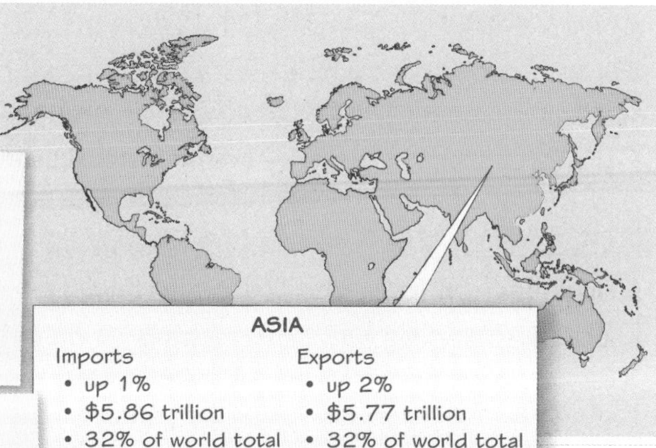

Countries can be significant for lots of reasons – for example, they might have world-famous cultural treasures, like the Great Wall of China. For development, a country's economic and political significance are very important.

This map has a trade summary for Asia. How does the location of your emerging country influence its trade with the world and its region?

Key facts

You need to know some key facts about what your country is like: what sort of political system it has, what its culture is like, what it is like in terms of physical geography.

India is an emerging country with a world-famous cinema industry. Bollywood makes 1600 films a year, seen in total by 2.7 billion people. What do you know about the culture of your emerging country?

You do not need to know this **contextual information** in huge detail. You'll be using this information to help explain *how your country is developing*.

Economic facts

You need to know key economic facts about your country, including how the following have changed since 1990:

- GDP and per capita **gross national income (GNI)**
- economic sectors (how they have changed)
- imports and exports
- **foreign direct investment (FDI)** – what sorts of TNCs are investing in your country.

India's GNI per capita (US$)

US dollars: 0, 200, 400, 600, 800, 1000, 1200, 1400, 1600, 1800

Years: 1990, 1991, 1992, 1993, 1994, 1995, 1996, 1997, 1998, 1999, 2000, 2001, 2002, 2003, 2004, 2005, 2006, 2007, 2008, 2009, 2010, 2011, 2012, 2013

http://www.indexmundi.com/facts/india/gni-per-capita

GNI per capita is usually measured in US dollars so that one country can be compared with another. It is GNI divided by the number of people in the country.

Make sure you know the percentage change in GDP and GNI per capita for your country between 1990 and now.

Now try this

Put together a fact file of information about your case study emerging country. Try to fill one side of A4 paper with information about its geography and key economic facts. Practise using these details in exam questions for this part of your course.

Globalisation and change

For your case study of an emerging country you need to know how globalisation and rapid economic change have affected your country.

The role of globalisation

Some emerging countries have seen very rapid economic development because of globalisation:

- transport technology – **containerisation** made global trade much cheaper
- internet technology – rapid, cheap **communication** between countries
- **TNCs** – leading the outsourcing of manufacturing and services to emerging countries
- government policy – emerging countries invested in infrastructure to attract TNCs, set up low-tax, low-regulation enterprise zones: encouraging FDI.

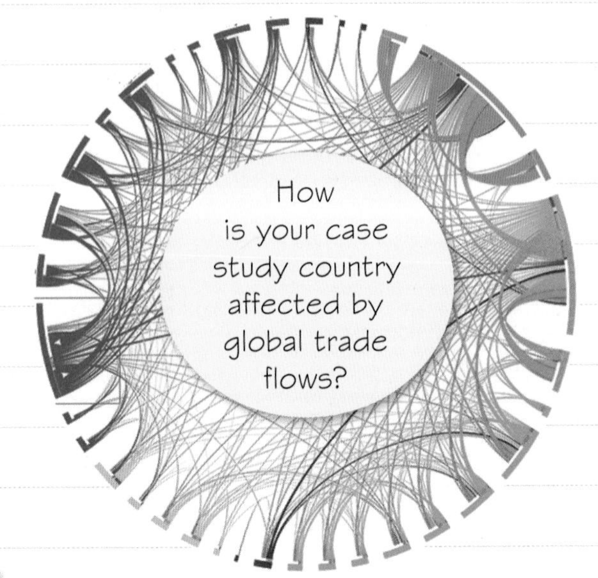

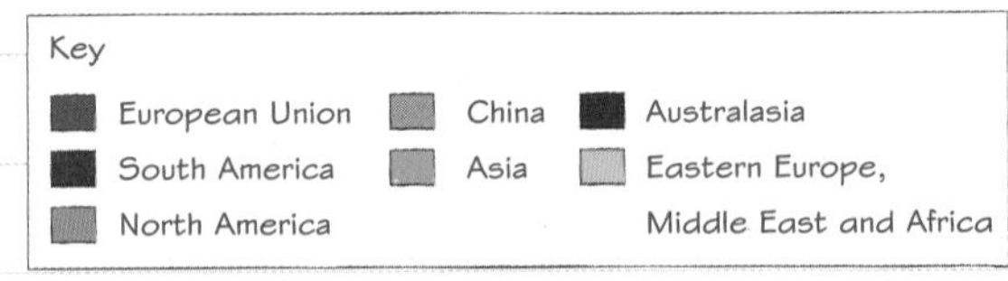

You should be able to describe how these globalisation processes affected the development of your emerging country.

Demographic change

The rapid economic development of emerging countries leads to the following:

- Rapid population growth.
- Improved medical technology and health education lower infant mortality.
- Fertility remains high during the country's rapid economic growth.
- Rising prosperity (wealth) and/or government policy starts to reduce fertility.
- Improved living standards and healthcare increase life expectancy.

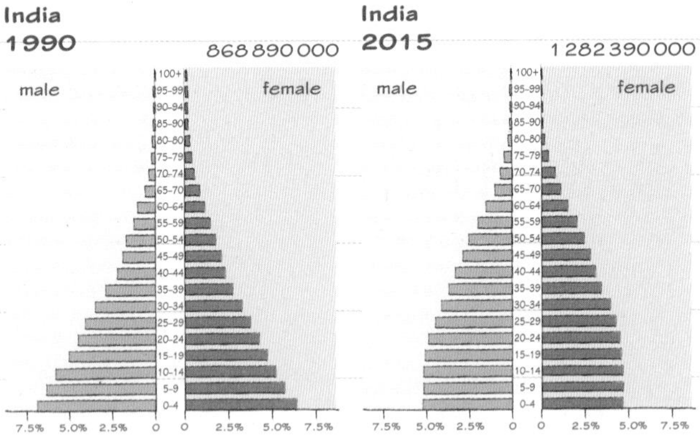

India 1990 — 868 890 000 — male / female

India 2015 — 1 282 390 000 — male / female

India's population has increased by 48 per cent since 1990. Do you know the population increase for your case study country?

Regional changes

Rural-urban migration has caused regional changes in many emerging countries – how about in yours?

- Rural–urban migration causes rapid city growth with young populations.
- Rural–urban migration leaves old people in the countryside.
- Regions remote from rapid industrialisation and urbanisation may be much poorer.

Now try this

Find a graph of population growth for your emerging case study country for the last 30 years or so (try the Index Mundi website). Sketch it or print it out (with room to annotate around it) and add as many labels as you can to describe the causes and consequences of this demographic change.

Economic development

Case study For your case study of an emerging country, you need to know about the positive and negative impacts of economic development and globalisation on people and on the environment.

Rural-urban migrants are mainly young men.

Economic development increases GNI per capita.

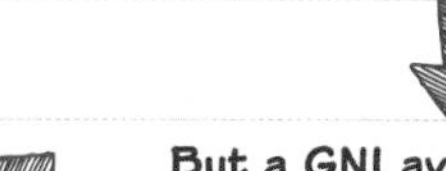

But a GNI average hides the growing inequality between rich and poor.

Women and old men are left in the countryside.

There are positive and negative impacts on different age groups and for men and women.

Positive impacts

- New jobs and skills (brought in by TNCs) especially benefit young migrants.
- Reduced poverty worldwide – in the last 20 years, nearly 1 billion people have been lifted out of extreme poverty (mostly in China).
- New technologies (e.g. 'green revolution') can reduce workload for rural women.

Negative impacts

- TNCs may decide to pull out of a region or country, causing unemployment.
- Economic development has increased inequality between the very rich and the very poor in many emerging countries.
- Pace of change is very rapid, leaving old people feeling lost and out of place.

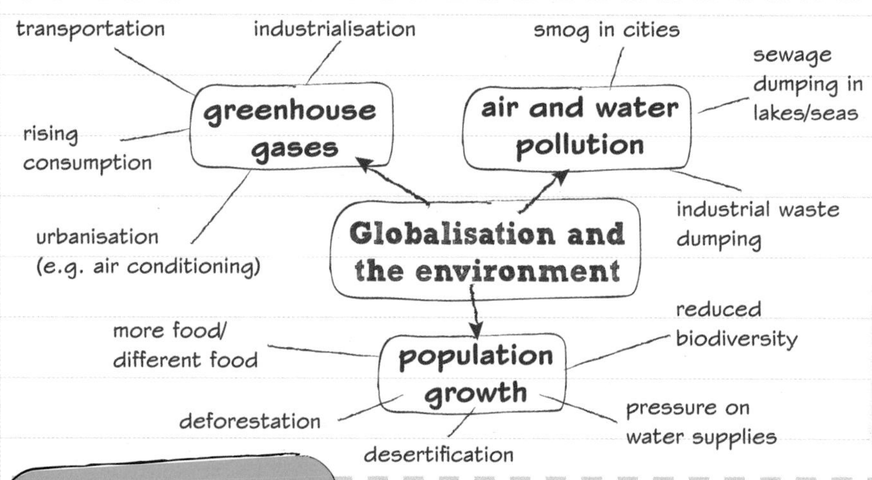

You need to know about the impacts of economic development and globalisation on the environment in your emerging country – for example, details on pollution and greenhouse gas emissions and how these affect human health (e.g. smog in big cities and increased death rates).

Now try this

Create a table with the title 'Economic development and globalisation' with two columns: Positive impacts and Negative impacts. Fill in specific examples from your case study notes. Two positive impacts and two negative impacts on people, and two positive impacts and two negative impacts on the environment would be ideal.

Had a look ☐ Nearly there ☐ Nailed it! ☐

International relationships

Case study For your emerging country case study you need to know how rapid economic development has changed the **geopolitical** influence of your country and its relationships with the EU and USA.

Definition

Geopolitical – the effects of geography on international politics

An example of geopolitics: Japan and China are rivals, so Japan has been keen to build strong economic relations with India, China's economically powerful neighbour in south Asia.

Regional influence

In the 20th century the world was dominated by two superpowers: the USA and USSR. In the 21st century global influence is split into regions – like Asia Pacific, or the EU, or Australasia. A country like Germany can have significant influence in the EU, but little influence in, say, Asia Pacific. The BRICS (Brazil, Russia, India, China and South Africa) all have lots of regional influence.

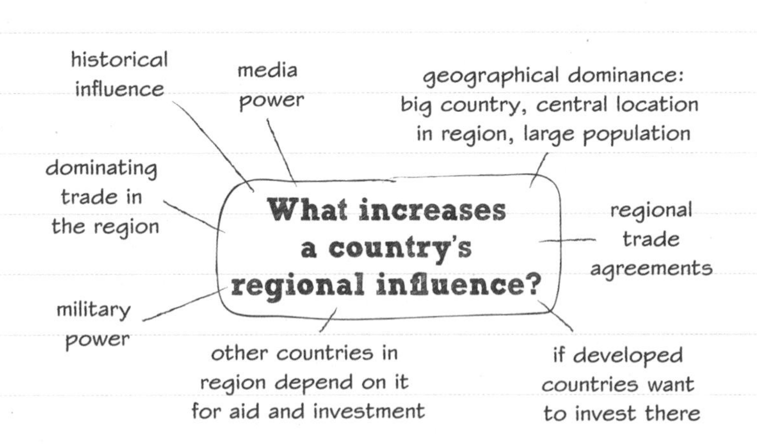

International organisations

You also need to know about your country's role in international organisations and its relationships with the EU and USA.

These pictures show meetings between the Indian prime minister and different foreign leaders. What are relations like between your country and the USA and EU?

What is your emerging country's role in organisations like the United Nations, World Trade Organization or G20?

Now try this

1 Find out the top three trading partners of your emerging country – the three countries it exports most to and the three countries it imports most from.
2 Why do other countries want to improve relations with rapidly developing emerging countries?

Costs and benefits

Case study For your emerging country case study you need to know about the costs and benefits of changing international relations. You also need to know why there are costs and benefits of having TNCs investing in your case study country.

Benefits of growing global importance

Emerging countries become more important mainly because of rapid economic growth.

- Other countries and TNCs want to invest in them so their investments grow rapidly too.
- As the emerging country gets richer, other countries want to sell products and services to them.
- Emerging countries often provide aid and investment in their region, which creates close ties between countries.
- Emerging countries often invest in military strength. They may play an important part in the defence of the whole region, or become important because they threaten other countries..

Costs of growing global importance

Changing international relations can cause some challenges for emerging countries.

- Developed countries block emerging countries from sharing global power.
- There is pressure for emerging countries to make cuts to greenhouse emissions.
- Emerging countries often come under international pressure about human rights issues within their country.
- Other neighbouring countries may become hostile as a result of military build-up in the emerging country.

Benefits of foreign investment

- FDI brings in a lot of money – for example, India received FDI of $24 billion in 2012, while China received $258 billion!
- TNCs bring big brands into the emerging country, which helps develop a bigger consumer market.
- TNCs often pay more than local companies, which pushes up wages.

In 2011 a major US soft drinks TNC announced an investment of $2 billion in India. What might the impact of this FDI be for Indian soft drinks manufacturers?

Costs of foreign investment

- The big brands brought in by TNCs can outsell local companies' products, leading to reliance on TNCs for goods.
- FDI is not always reliable. India lost $14 billion of FDI after the global crash of 2008, when TNCs cut back on foreign direct investment.
- Lack of regulation of TNC activities can have environmental consequences.

Key things to know about your case study

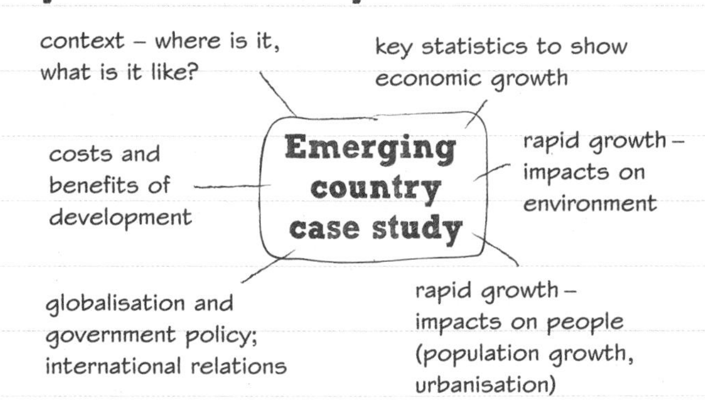

Now try this

Use the diagram on this page to check that you have all the information you need to revise your emerging country case study properly.

 Had a look ☐ Nearly there ☐ Nailed it! ☐

Urbanisation trends

The world's population is becoming increasingly urbanised. Cities in developing global regions are growing especially fast.

Developed regions

Europe, North America, Australia New Zealand and Japan

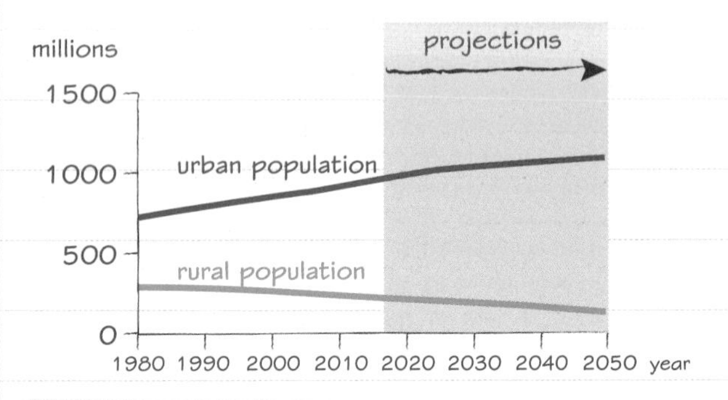

Developing global regions have very rapid urbanisation. The rate of urbanisation in developed global regions is much slower. This is mainly because developed countries are already highly urbanised. For example, 80 per cent of people in the UK live in cities.

Developing regions

Africa, Asia (excluding Japan), Latin America and the Caribbean, Melanesia, Micronesia and Polynesia

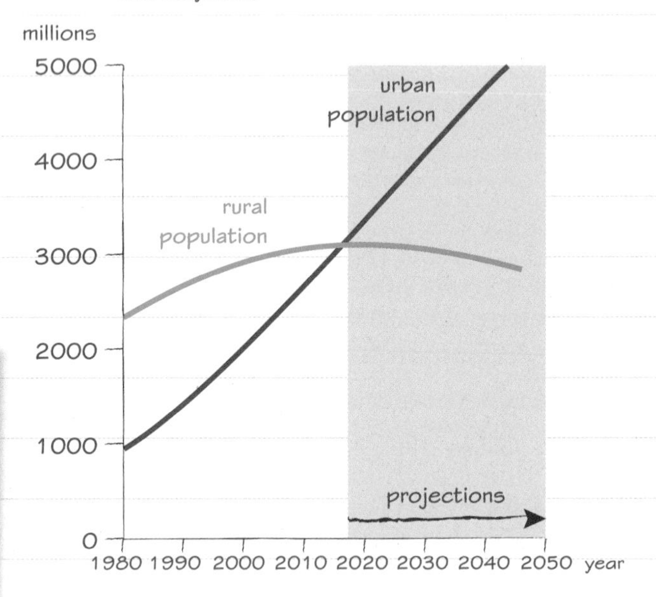

Worked example

Explain why most of the future growth of cities is likely to be in the developing world. **(4 marks)**

Cities in developing countries are growing much faster than cities in developed countries. In developing countries, the urban population is growing because of natural increase – the birth rate is higher than the death rate. There is also migration from rural areas to urban areas (urbanisation). Many people are moving to the cities because agricultural practices are becoming more mechanised and there are fewer jobs, so people move to cities for work. Drought or other natural disasters may also force a move. Generally, urban areas provide people with better living conditions, such as piped water, electricity and healthcare, and higher wages.

Calculating the rate of change

You may be asked to calculate the **rate of change of urbanisation**.

You need to divide the actual increase by the original value and then multiply it by 100.

Percentage increase =

$$\frac{\text{actual increase}}{\text{original value}} \times 100\%$$

Now try this

The urban population of the world in 2000 was 2.84 billion. By 2005 it had risen to 3.15 billion. What was the percentage rate of change from 2000 to 2005? Round your answer to the nearest whole number. **(1 mark)**

Hint: The actual increase is 3.15 billion minus 2.84 billion, and the original value to divide it by is 2.84 billion.

Megacities

You need to know about megacity sizes, locations and growth rates. Some megacities cause problems of '**urban primacy**': they 'suck' development out of the rest of the country.

Definitions

Megacity – a city with at least 10 million inhabitants

World city – a city with a dominant role in global processes

Urban primacy – the most important city in a country, which dominates the rest of the country

Hinterland – the region around a city

Top five megacities

There were 35 megacities in 2015. This table shows the five largest cities in that year.

Megacity	Country	Size (million)
1 Tokyo-Yokohama	Japan	37.8
2 Jakarta	Indonesia	30.5
3 Delhi	India	24.9
4 Manila	Philippines	24.1
5 Seoul-Incheon	S. Korea	23.4

Megacity locations, 2015

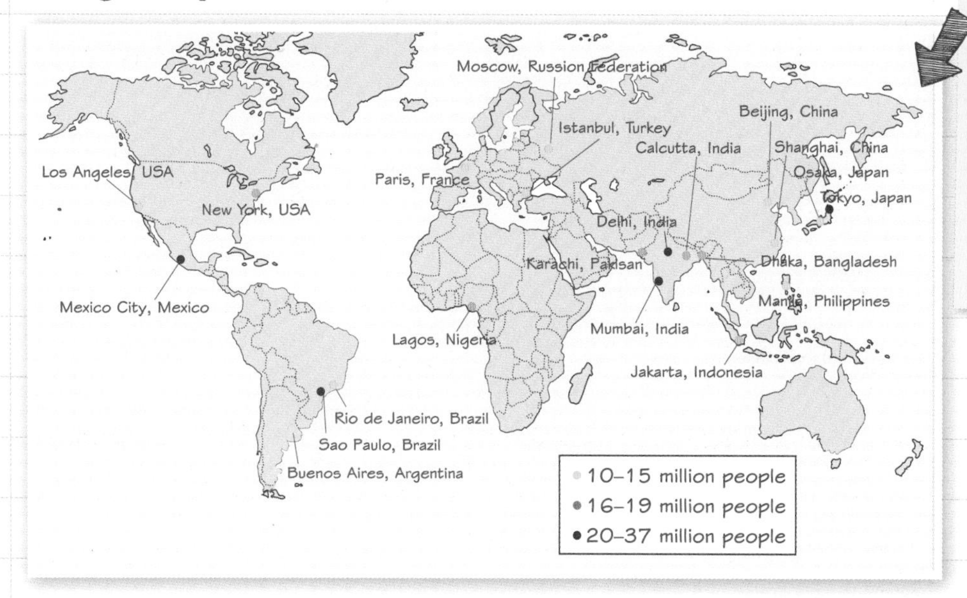

Location: Asia is the centre of megacity growth. By 2025, Asia should have at least 28 megacities. Megacities are created where economic development is rapid.

Growth rate

In 1950 there were only two megacities: New York and Tokyo. By 1985 there were nine. Megacities in developing and emerging countries are growing very fast. For example, Mumbai's population doubled in size between 1991 and 2013. Growth rates are fuelled by rural–urban migration.

Worked example

Remember that 'explain' means 'say why.'

Explain **two** reasons why urban primacy can become a problem for developing and emerging countries. **(4 marks)**

The main problem of excessive urban primacy is that all economic growth is concentrated in the city while the city's **hinterland** is deprived of growth.

A second problem is that political power also becomes focused in the city. Decision making that affects the whole country takes place in the primate city and is skewed toward the city's needs, not the country's.

Now try this

In 2015, London had a population of 8.6 million. Calculate the percentage increase required for London to become a megacity. Round your answer up to one decimal place. **(2 marks)**

Urbanisation processes

Economic change and migration are key factors in why cities grow or decline. These processes in developing and emerging countries have some similarities with developed countries, and some differences.

Definitions

Rural–urban migration – when people change where they live from rural areas to urban areas

International migration – when people move to live in another country

Internal migration – when people move from one part of a country to live somewhere else within that same country

Natural increase – the difference between the number of births and deaths in a year

How do cities grow (or decline)?

There are two main ways:

- migration – people move to live in the city, or move away from the city
- **natural increase**

These two processes are connected. It is usually young people who migrate to live in cities. The birth rate for young people is higher than the death rate for old people.

Economic development creates a multiplier effect

- economic growth means people move to the city
- more people = bigger workforce so more industries
- more industries = more jobs
- more people = more customers
- more people = more houses to be built

Rural-urban migration is the main cause of rapid urbanisation in emerging and developing countries

Rural	Urban
few opportunities	many more jobs
poor healthcare	better paid
low pay, difficult work	better education
low level of education	better healthcare
	better housing
	modern lifestyle
	many more opportunities

Worked example

Study the photo opposite.

Explain **two** reasons why economic changes can cause a city to decline in population. **(4 marks)**

Economic problems can mean that industries shut down. If it is no longer easy to get a job in a city, people will move away. City governments rely on taxes from businesses and residents to make the city a pleasant, safe place to live and work. If economic problems mean that tax payments go down, the city starts to become run-down, crime rates rise and more people move away.

An abandoned building in the US city of Detroit, which has lost over 60 per cent of its population since 1950.

Now try this

In 2015 London's population reached 8.6 million – the same size it was in 1939. Suggest **one** reason why London's population declined after 1939 and **one** reason why it has grown again. **(2 marks)**

Differing urban economies

You need to know why urban economies are different in developing, emerging and developed countries. Key differences are in formal and informal employment, in the importance of different economic sectors and in working conditions.

Definitions

Formal employment – jobs that pay taxes and provide workers with job security and legal protection (e.g. Health and Safety laws). These jobs are hard to get in developing countries.

Informal employment – jobs that are not regulated: informal workers pay no taxes but are not protected by law. These jobs are easy for new arrivals in a city to get.

Working conditions – in emerging and developing countries there may be unsafe, stressful or uncomfortable working conditions (e.g. hot, polluted); long working hours without enough breaks; harsh penalties for lateness or absence.

Street hawkers in Lagos, Nigeria. This is **informal employment**. The hawkers don't pay taxes and have no legal protection. Informal employment is characteristic of developing countries.

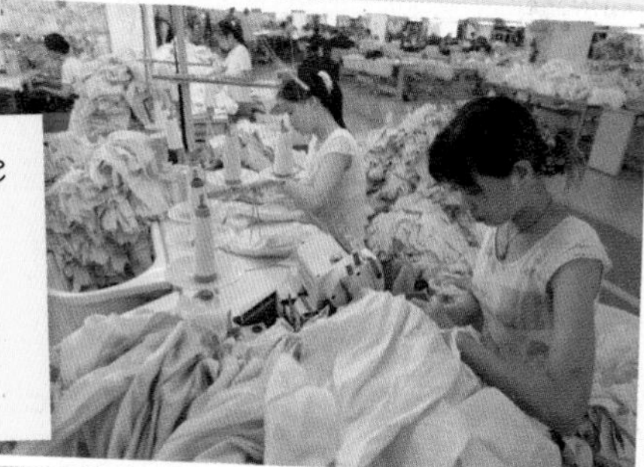

Workers in a clothing factory in Shenzhen, China. The **urban economies** of emerging countries have much more manufacturing than the urban economies of most developed countries. **Working conditions** are often tougher than in developed countries.

Worked example

Contribution of economic sectors to Mumbai's economy (per cent of total)

Years	Primary (%)	Secondary (%)	Tertiary (%)
1993–4	1.25	36.12	62.64
2005–6	0.88	25.30	73.82

Study the table above. Which one of the following does it suggest would have grown the most in Mumbai between 1993 and 2006? **(1 mark)**

☐ textiles manufacturing ☐ farming
☒ IT services ☐ steel production

The tertiary sector is the service sector so IT services is the correct choice here.

The economic sectors of developed countries' cities are often dominated by services, including tourism. This is the British Museum – visited by 6.8 million people in 2015.

Now try this

State **one** advantage of informal employment in cities in developing countries. **(1 mark)**

Had a look ☐ Nearly there ☐ Nailed it! ☐

Changing cities

Cities change over time, and these changes show up in the way land is used in the city.

Definitions

Urbanisation – an increase in the number of people living in cities

Suburbanisation – the movement of people, industry and jobs from the centre of the city to its outer areas

Counter-urbanisation – the movement of people out of the cities into the countryside (developed world process)

Regeneration – new investment into old, run-down parts of the city (often inner city areas)

Factors affecting land use

You can tell the area of a city where land has highest value, because that's where the tallest buildings are.

Change over time

Theories of city development suggest these stages.

1. The **central business district (CBD)** is located where the city first developed, where all the major roads join.
2. A manufacturing zone develops.
3. New migrants live in this inner city zone, where housing is poor but cheap and they are close to their jobs.
4. Developing public transport lets richer people live further out, in the pleasant suburbs.
5. The city gets too congested for industry, which moves out to cheaper land in the suburbs.
6. The inner city areas get poorer.
7. As the city expands, commuter journeys from the outer suburbs become very long. Some wealthier residents move back to the inner city and redevelop the old housing.

Worked example

Study the satellite image of the city. Describe **two** features of the land use you can see in this image. **(2 marks)**

There is an industrial zone in the northern half of the city, located next to the sea. The industrial units are widely spaced out on flat land. In the south of the city the land use is residential plots, suggested by the small size of the blocks and the green gardens and parks.

Now try this

Which of the following factors affecting city land use would best explain why a polluting industry has located away from residential areas of a city? **(1 mark)**

☐ **A** accessibility ☐ **B** availability ☐ **C** cost ☐ **D** planning regulations

Location and structure

Case study You will have studied a megacity in an emerging or developing country. We've used the example of Mumbai, in India, here to illustrate the sort of information you need to know about your case study. In the exam, refer to the case study you did in class.

Make sure you know basic geographical information about your megacity case study.

The location of your megacity within its country and in the world has influenced its growth into a megacity. Where it is will also have affected its structure and the economic sectors that are important to it.

Mumbai has one of the deepest natural harbours in the world. This gives it big advantages as a major port on the Arabian Sea.

Mumbai was built on seven islands. Space for city growth was highly constrained (limited), making land very expensive. New Mumbai was developed on the mainland from 1971.

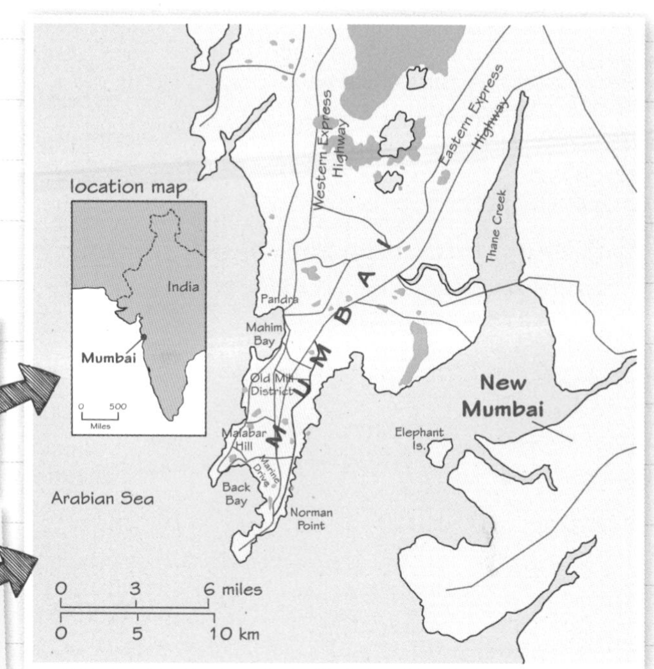

Megacity structure

You need to know about the structure of your megacity: its CBD, inner city, suburbs, urban–rural fringe. Have an idea about when each of these developed and what they were used for.

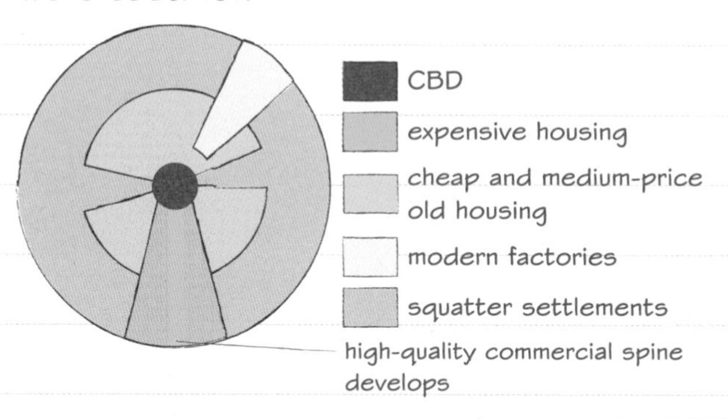

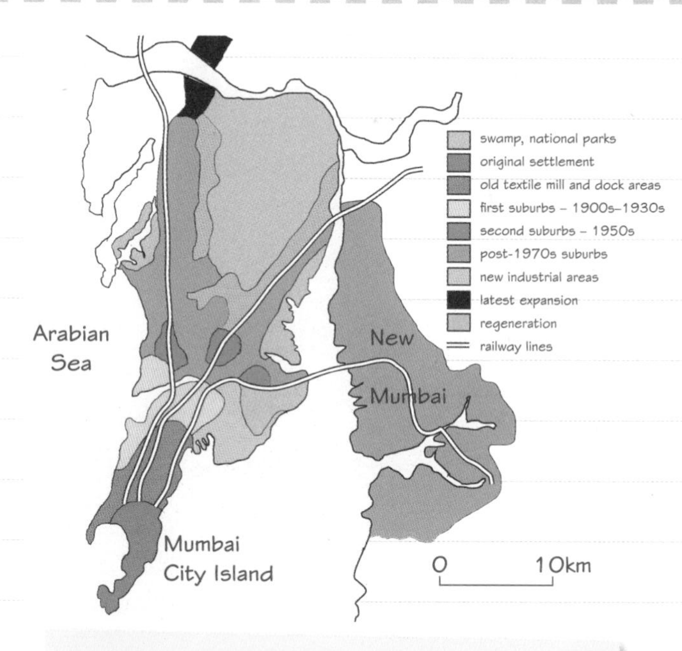

The structure of megacities in emerging or developing countries is often quite complex. Transport routes are often important.

Mumbai has had several stages of expansion, with suburbs spreading along railway routes and onto the mainland. Industrial sectors have moved away from the old city centre too.

Now try this

Use Google Earth (you can access it from Google Maps or Google Street View by clicking on the panel called Earth) to locate your megacity. See which different areas you can locate by zooming in and looking at the different types of building. Locate: the CBD, slum areas or worker housing areas, suburban areas, and industrial areas.

Megacity growth

Case study Why has your emerging or developing country megacity become a megacity (population of 10 million or over)? What were the factors that caused its population to grow? What makes people want to live there? We've used Mumbai as an example here: revise the megacity you did in class.

For your megacity case study, you need to know:

- how its population grew (some statistics)
- reasons for its population growth.

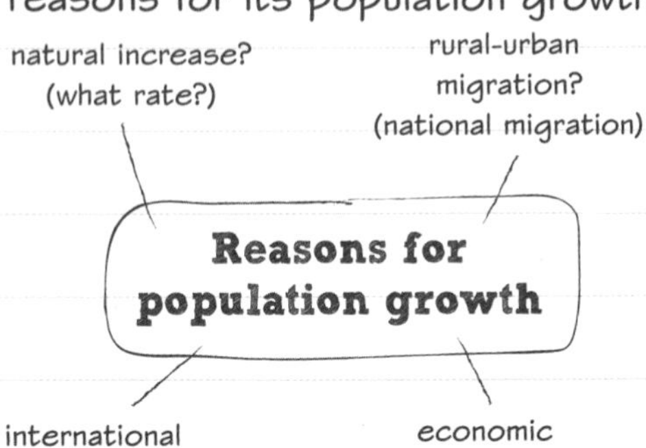

Mumbai fact file

- Estimated population 2013: 12 million people
- 1991 population: 9.9 million
- Population growth is 2.9 per cent per year
- Size of Greater Mumbai city area: 603 km^2
- Population density: 20482 people per km^2
- GDP in 2015 was US$278 billion
- 42 per cent of people live in slums (2010)
- 68 per cent work in the informal sector (2013)

Now try this

Put together a fact file like this one for your megacity.

You need to know how population growth has affected the way land is used in your city and the ways your city has grown.

Mumbai: reasons for growth

- Rural-urban migration – 1000 national migrants arrive in Mumbai daily, 9 out of 10 from rural areas.
- High rate of natural increase – Indians still like to have larger families although the birth rate is falling.
- City has strong economy – foreign investment from the 1990s saw many services outsourced to Mumbai from developed countries. Now it is the centre of hi-tech industries, Bollywood and finance.
- Huge informal sector – most people in Mumbai are poor and live in slums.

You need to know about the **opportunities** for people living in your megacity. What makes them want to live there?

- Access to jobs (often to send money back home to their families)
- Job promotions or job transfers
- Access to better education for their children
- Access to better healthcare
- Better marriage opportunities
- More entertainment options

Reasons for moving to Mumbai

	Rural–urban (%)	Urban–urban (%)
To find a job	68.4	47.2
To start a business	5.5	7.6
Education	2.8	4.9
Marriage	8.0	17.4
Job transfer	3.4	6.9

Rural–urban and urban–urban migrants may have different reasons for moving.

Megacity challenges

Rapid population growth has caused major challenges for many of the people living in megacities in emerging or developing countries.

Housing

Rapid population growth means there are far more people arriving in the city than there are affordable houses for them to live in.

That shortage of affordable housing means people are forced to live in:

- slum housing – often with many people sharing each room
- shanty towns (squatter settlements) – where people build housing out of whatever materials they can find.

Challenges of slums and shanty towns

Slums and shanty towns

- homes built from scrap materials on any spare land
- dangers from fire, flooding and landslide
- there is no clean water, electricity, rubbish collection or organised sewage disposal
- life in the shanty towns is very stressful
- crime rates are high
- litter and sewage create a breeding ground for disease
- people are malnourished because there is a lack of money and food

Water supply and waste removal

Squatter settlements often do not have a piped water supply, at least when the settlement is first developing.

- People can buy bottled water to drink, but it is expensive and not everyone can afford it.
- Some people take water from nearby rivers or streams, which are often polluted and carry diseases.
- People often go to the toilet in waste ground, disposing of their excrement (poo) in plastic bags left there. These waste areas smell bad and diseases can spread.

A toilet in a Mumbai slum. The toilet is not connected to a sewerage system.

Employment and opportunities

In megacities like Mumbai, most people work in the informal sector.

- Pay is low and not always regular or reliable.
- Working conditions can be dangerous.
- There are no benefits or security: if people get ill and can't work, they don't get any money.
- As pay is low, many children are put to work instead of going to school. This limits their opportunities in life.

This street barber in Mumbai is part of the informal economy. For poor people the range of services on offer is small and very basic. This barber has basic facilities but will charge an affordable price for his services.

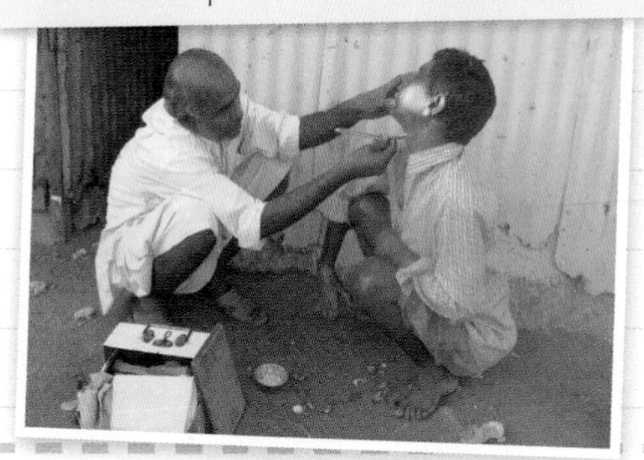

Traffic congestion in megacities is often legendary! The desire for moped and car ownership is very high, but that means the streets are clogged with traffic.

Now try this

Put together a table headed 'Megacity challenges', with columns for 'Housing', 'Water and sanitation', 'Employment conditions', 'Service provision' and 'Traffic congestion'. Add at least **one** relevant detail from your megacity case study to each column.

Megacity living

Case study

Megacities in emerging and developing countries can have very wealthy areas right next to very poor areas. This can pose major political and economic challenges for city managers.

Dharavi, a slum area of Mumbai. Where are the areas of slum or squatter settlements in your megacity?

Andheri is a popular suburb for middle-class people in Mumbai. Andheri railway station is one of the busiest stations in the world because of people commuting to work. Where are the wealthier areas in your megacity?

The new suburb of New Mumbai. This suburb is growing very rapidly as middle-class people leave congested Mumbai to go to the mainland. What quality-of-life benefits do new suburbs provide?

These three photos show scenes from Mumbai. Mumbai has both Asia's largest slum and the world's most expensive single-family home!

Squatter settlements often develop close to the city centre because the poorest people cannot afford to travel far to work. However, this causes **political** difficulties for city managers as the settlements can be on valuable land. There is a choice between moving the inhabitants to new areas, or helping them to develop their settlement.

Middle-class people commute to work in the city from suburban locations. This causes **economic** problems for city managers with car and moped congestion on the roads that slows the city's economy down, leading to air pollution and commuter stresses that impact on people's quality of life. Mumbai commuters suffer 'super-dense crushloads' in packed trains and buses.

Squatter settlements are usually illegal and so city managers have legal rights to clear the land of houses. However, this is very problematic politically. If poor people are rehoused on cheap land at the edge of the city, there are often no services and no affordable way to reach work.

Now try this

Use Google Maps to locate a wealthy area and a poor area in your megacity. Provide at least one reason to explain the location of each.

Megacity management

There are different strategies for tackling the challenges linked to rapidly rising megacity populations. Each has advantages and disadvantages.

Definitions

Sustainability – in cities, this means planning for an efficient city that uses less energy and minimises pollution to create higher living standards for all.

Top-down strategies – large-scale ways of improving city sustainability that are planned, funded and managed by the city government.

Bottom-up strategies – small-scale ways of improving city sustainability that do not involve governments directly. They are often funded by NGOs or community organisations.

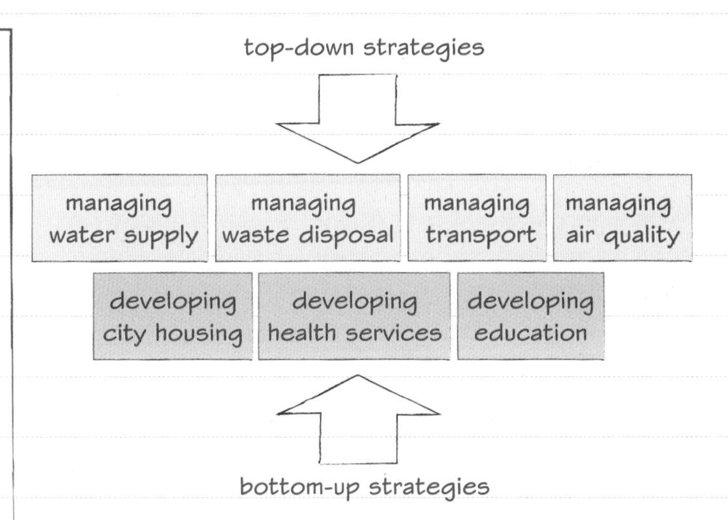

Top-down advantages

- Large-scale, so the problems of the whole city can be tackled together.
- City government has the political power necessary to make landowners sell their land for new developments.
- Governments can pass laws to change people's behaviours – for example, to stop industries dumping waste in city rivers.

Top-down disadvantages

- Impose changes on people that they may not like – for example, slum clearance.
- City governments can become biased – for example, toward the needs of big business, looking after them instead of working for poorer people in the city.
- Top-down strategies are expensive and complicated, so can end up going over budget, adding extra taxes for residents.

Bottom-up advantages

- Target specific needs of particular local communities – for example, health workers to visit slum areas, provision of affordable public toilets.
- Some city governments (e.g. Mumbai) refuse to recognise slums as legal, so the only help people living there get is from NGOs and community initiatives.
- Can have positive **multiplier effects** – for example, improving child health means children can attend school more.

Bottom-up disadvantages

- Cannot fix the city-wide problems like transport congestion.
- City governments that should be fixing poor slum conditions may leave it to NGOs and local communities to fix the problems.
- By helping slum communities to improve services, schemes can cause conflict with city governments that want to clear sites for more profitable land uses.

Now try this

For your case study megacity: a) identify at least **two** advantages and disadvantages of top-down strategies and b) identify at least **two** advantages and disadvantages of bottom-up strategies.

Paper 1

Paper 1 is Global Geographical Issues. This paper has three sections: A, B and C. Each section has a range of different types of question, including an 8-mark extended writing question.

The three sections of Paper 1 are:

- A: Hazardous Earth
- B: Development dynamics
- C: Challenges of an urbanising world

You can expect to find one 8-mark question at the end of each section. Here are some examples.

> 'Being able to accurately forecast tropical cyclones is the best defence a country can have against these hazards.' Assess this statement. **(8 marks)**

> For a named emerging country, assess the importance of globalisation in its development. **(8 marks)**

SPGST

What is **SPGST**? It stands for:

> **S**pelling, **P**unctuation, **G**rammar and use of **S**pecialist **T**erminology

Your extended writing question in Paper 1 Section B will be worth 12 marks in total. That is because added to the 8 marks are 4 marks for SPGST. To get all four SPGST marks your answers need:

- accurate spelling and punctuation
- to follow the rules of grammar
- a wide range of specialist terms used appropriately.

A game of two halves

The 8-mark questions involve the command words 'Assess' or 'Evaluate'.

These questions are worth 4 + 4 marks.

- 4 marks are for Assessment Objective 2
- 4 marks are for Assessment Objective 3

Assessment Objective 2 is about how you use your **knowledge of geographical concepts** and places, environments and processes in your answer.

Assessment Objective 3 is about **applying** your knowledge and understanding to geographical issues, using evidence to come to a judgement.

What to aim for

These extended writing questions are marked by levels. A top answer will show you:

- **understand** the geographical concepts involved in the question
- **know how** processes, places and the environment are linked
- **can apply your understanding** to unpick the different factors involved in the question
- **can put together a clear, logical argument**
- **can use evidence** to decide which of the factors are more important than others.

Extended writing questions

Study this 8-mark Section A question:

> 'Human activity makes current global warming completely different from past climate change.' Assess this statement. **(8 marks)**

What is it asking you to do?

1. Show your understanding of the causes of past climate change.
2. Show your understanding of the causes of current global warming.
3. Consider the differences between them.
4. Consider the similarities between them.
5. Make a judgement: similar or different?

Now try this

Try this global warming question for yourself using the information on this page to help you.

Uplands and lowlands

Geology and past processes – such as glaciation and past tectonic activity – have influenced the physical landscape of the UK.

There are three groups of rock type.

- **Igneous** – made from magma (granite)
- **Sedimentary** – compressed sediment (e.g. clay, chalk, limestone)
- **Metamorphic** – igneous or sedimentary rock changed by heat or pressure (e.g. shale into slate)

The UK is split into two halves geologically.

- The geology of the top half is mainly igneous and metamorphic rocks. This forms **upland** landscapes.
- The geology of the bottom half is mainly sedimentary rocks. These rocks are characteristic of **lowland** landscapes.

Glaciation

The top half of the UK was glaciated during the last Ice Age. Ice sheets and glaciers hundreds of metres thick covered the land as far south as London. The ice pressed down on the landscape and eroded it in distinctive ways.

The bottom half of the UK was not covered in ice sheets, but it was heavily influenced by glacial deposition. Clays, sands and silts eroded by glaciers in northern areas were dumped and washed over southern areas. The south was frozen, even if it was not ice-covered.

Worked example

Study the picture opposite, which shows Arthur's Seat, a long-extinct volcano near Edinburgh. Explain **two** ways in which the UK's landscape has been influenced by past tectonic processes. **(4 marks)**

North of the Tees-Exe line the UK's geology is largely igneous: rock formed from magma, associated with tectonic events. Long-extinct volcanoes also form other hills and mountains in the UK.

This is a good answer because it relates tectonic events to the UK landscape.

Now try this

Draw an annotated sketch of **one** of the following UK landscape features: a U-shaped valley (upland), a meandering river (lowland), a chalk coastal cliff and stack, a clay cliff slumping into the sea. **(3 marks)**

Main UK rock types

The UK's main types of igneous, metamorphic and sedimentary rocks help produce some characteristic UK landscapes.

Main UK rock types

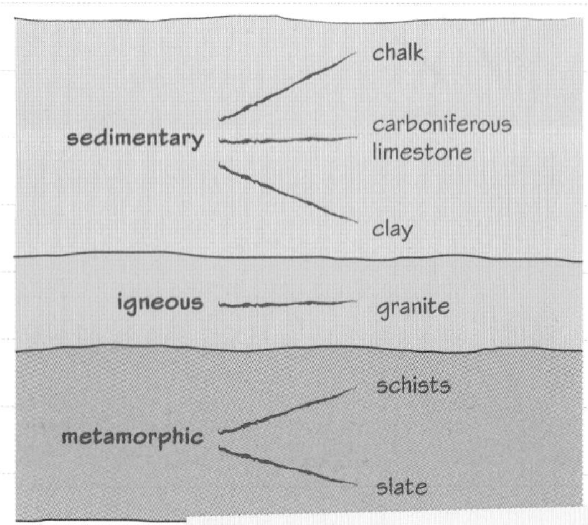

Sedimentary, igneous and metamorphic rocks in the UK

Chalk and clay landscapes

- Chalk is strong and **permeable** – water moves through it. It forms cliffs when it occurs at coastlines.
- Chalk is only found in lowland Britain.
- Clay is weak and **impermeable** – water cannot move through it.
- Clay is found all over Britain. Clay landscapes are typically wide, flat plains with lots of lakes, streams and rivers.

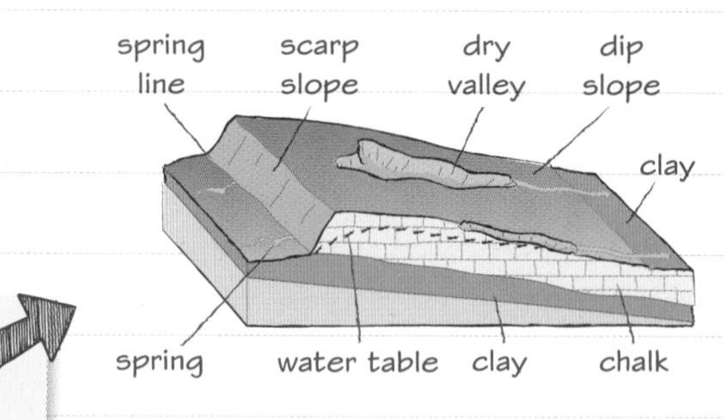

Water draining through the chalk flows out as springs along the line where the permeable chalk meets **impermeable** clay.

Igneous and metamorphic rocks

- Granite is hard and resistant to erosion but is susceptible to chemical weathering.
- Granite is impermeable and granite landscapes are badly drained – boggy.
- Tors are features of some granite landscapes: towers of granite chemically weathered into blocks.
- Metamorphic rocks are very strong and very resistant to erosion and weathering.
- Slate is formed from clay. Layers in the original clay form weak planes in the slate.
- Schists are formed from shale. The word schist originally meant 'to split'. Schist rocks split easily.

Worked example

Study the diagram below, which shows a cross section of a **carboniferous** limestone landscape. W is a limestone pavement, Y is a cave, and Z is a gorge. Which **one** of the remaining letters indicates a resurgence stream? **(1 mark)**

X

Carboniferous limestone is permeable and is chemically weathered by rainwater. Limestone landscapes have distinctive features, and are often associated with underground caves.

Now try this

Identify the rock type that underlies Dartmoor – an upland area with thin, boggy, poorly drained soils and dramatic tor features. **(1 mark)**

Physical processes

Some distinctive upland and lowland UK landscapes result from different physical processes working together.

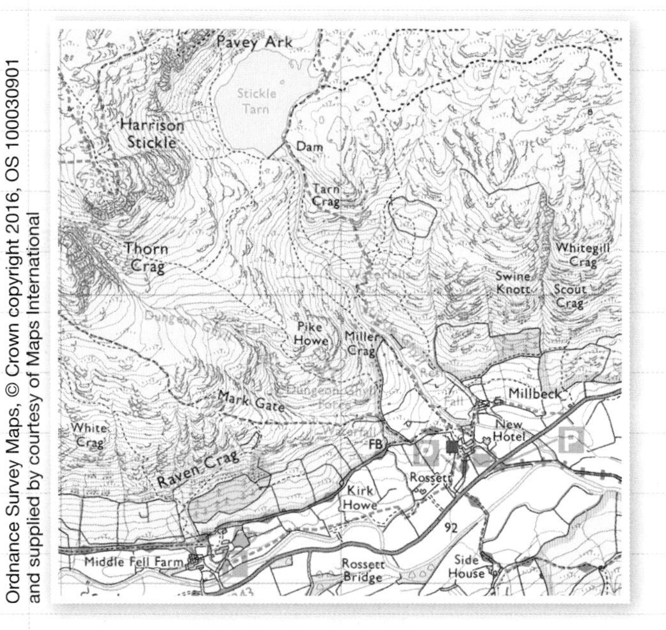

Upland landscapes

This upland landscape is in the Lake District. The (OS) map extract scale is 1:25 000.

- Stickle Tarn, at the top of the map extract, is a 'post-glacial' feature. It is where a glacier formed during the Ice Age creating a corrie.
- The crags in the map extract are exposed rock faces. Weathering of the rock leads to rock fragments breaking off and falling to the base of the cliff to form a scree slope.
- The high precipitation in the Lake District means there is a lot of surface drainage over the impermeable rocks – lots of streams.
- The valley floor at the bottom of the map extract is too wide for the stream that is in it. The flat bottom and steep sides show that the U-shaped valley was formed by a glacier.

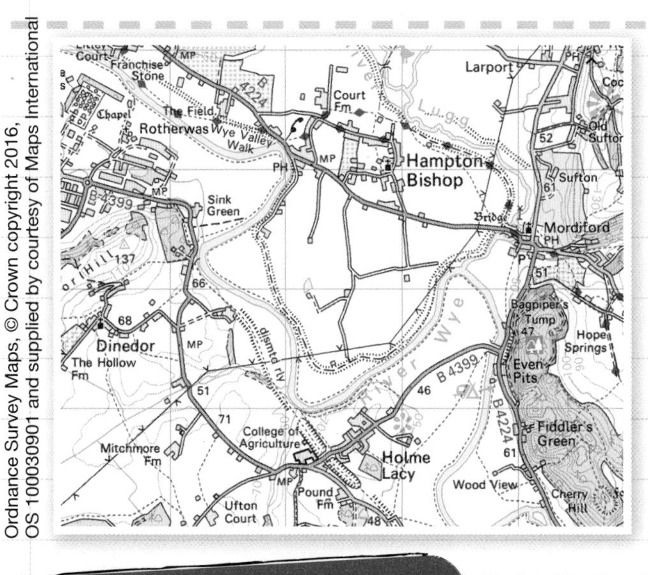

Lowland landscapes

This lowland landscape is in Herefordshire. The OS map extract scale is 1:50000.

- The landscape has been formed by the actions of two rivers: the River Lugg and the River Wye.
- As the rivers have meandered, they have eroded a wide valley between low hills.
- The rivers transport silt eroded from the river channel.
- When there is prolonged heavy rain in the region, the rivers flood and water spreads out all over the valley floor, depositing the silt to form a wide, flat **floodplain**.

Worked example

Study the photo opposite of a limestone pavement near Malham in the Yorkshire Dales. Bedding planes in carboniferous rocks are widened as rainwater, which is weakly acidic, reacts with the limestone, slowly dissolving it. Which of the following is the name of this form of weathering? **(1 mark)**

☐ **A** mechanical weathering
☐ **B** sub-aerial weathering
☐ **C** biological weathering
☒ **D** chemical weathering

Now try this

Which of the following slope processes occurs when the top layer of a slope made up of permeable material overlying impermeable material becomes saturated with rainwater, and a mass of soil and rock slips suddenly down the slope? **A** slumping **B** rock fall **C** soil creep **D** debris flow. **(1 mark)**

Had a look ☐ Nearly there ☐ Nailed it! ☐

Human activity

The UK has been settled by humans for many thousands of years and all its landscapes have been heavily influenced by human activity.

Agriculture

This OS map extract is at 1:25 000 scale. It shows a region of Suffolk in the east of the UK.

- The blue lines are drainage ditches, built to drain water away from low-lying agricultural land to allow crops to grow.
- Trees have been cleared to make way for agriculture.
- Straight lines on maps are not often produced by natural physical processes so they are a good indication of human activity.

Humans have altered almost all the landscapes of the UK through farming. Different farming types are suited to different landscapes – for example, sheep farming in upland areas and arable farming in fertile lowland valleys.

Forestry

Forestry is planting, managing and caring for forests for different purposes such as nature conservation, landscaping, recreation and timber production.

- Many UK upland landscapes have been planted with trees. Sometimes they are in straight rows to make forestry processes easier to manage.
- The UK would naturally be covered by deciduous woodland. However, some UK landscapes feature conifer plantations, which have been planted for timber production and are very distinctive.

When trees are felled for timber, a section of the plantation may be cleared. The aerial photo above shows a cleared area in the Strathyre Forest, Scotland.

Settlements

Settlements grew up where the landscape offered particular advantages. For example:

- river meander loops made good defensive locations
- natural harbours were sites for fishing villages
- shallow points of rivers were used as fords
- springs gave people reliable freshwater.

As settlements grew, the settlements took over the landscape. In big cities, many streams and small rivers now run in tunnels underground.

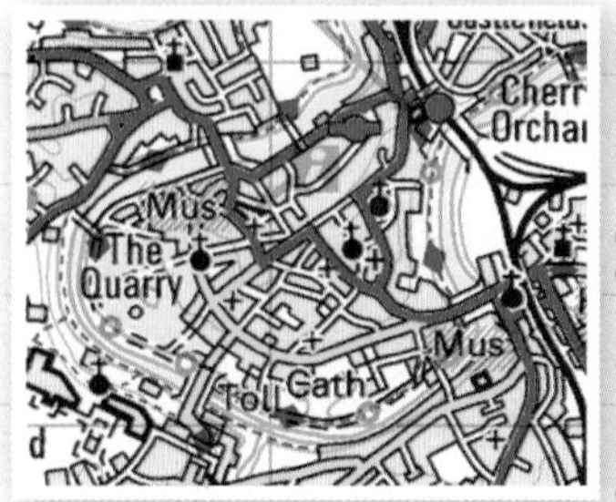

Now try this

The map above shows the Shropshire town of Shrewsbury. Suggest one reason why the site of Shrewsbury was chosen for a settlement. **(2 marks)**

Geology of coasts

The geological structure of coasts and the type of rock found there influence the erosion landscapes formed.

Soft rock (e.g. clay)

- Soft rock is easily eroded by the sea.
- Cliffs will be less rugged and less steep than hard rock coasts.
- Soft rock landscapes include bays.

Hard rock (e.g. granite)

- Hard rock is resistant to all types of erosion.
- Cliffs will be high, steep and rugged.
- Hard rock landscapes include wave-cut platforms and headlands where caves, arches and stacks are formed.

Concordant and discordant coasts

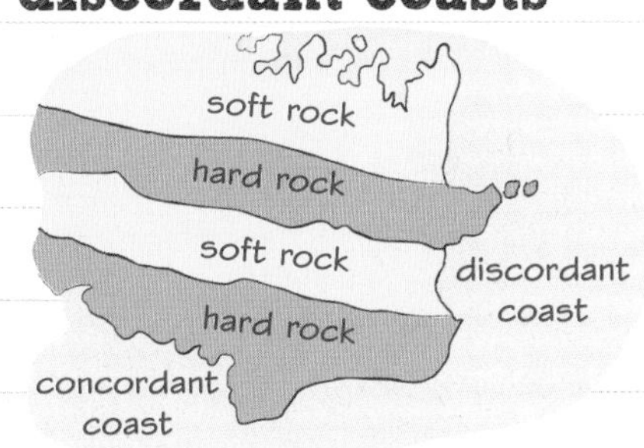

Concordant coasts are made up of the same rock type, parallel to the sea. On discordant coasts the rock type alternates in layers perpendicular to the sea, forming headlands and bays.

The hard rocks in this diagram are chalk and limestone; the soft rocks are mudstone, sands and clays.

Joints and faults

- Joints are small cracks in rock, and faults are larger cracks in rock.
- Both make rock more susceptible to erosion.
- Rocks with more joints and faults are eroded more quickly than rocks with fewer joints and faults.

Rates of erosion: other factors

How fast coastal erosion occurs is primarily down to geology (soft rock is eroded much faster than hard rock) but is also influenced by:

- geological structure – for example, if soft rocks and hard rocks occur together
- 'wave climate' – how powerful waves are, wave direction, wave height, fetch (how far winds travel over open water), etc.
- local currents and tidal range (the difference in height between low and high tides)
- groundwater levels – saturated cliffs (high groundwater) are more vulnerable.

Worked example

Study the photo below, which shows a chalk headland and stacks. Which point on the geological map opposite marks the location of this feature? **(1 mark)**

C

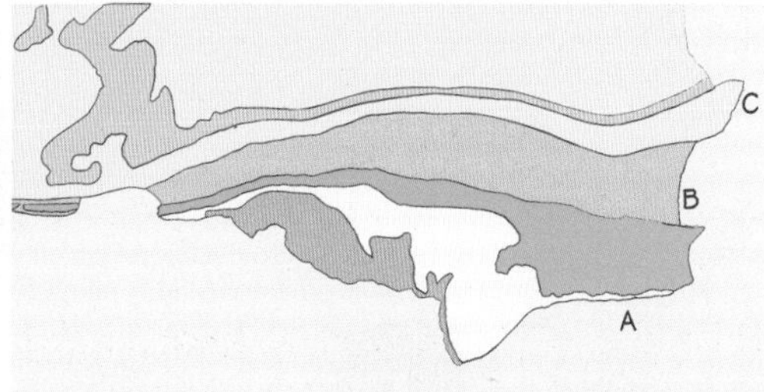

Now try this

Identify two landforms that are characteristic of a discordant coast.
(2 marks)

Had a look ☐ Nearly there ☐ Nailed it! ☐

Landscapes of erosion

You need to be able to explain how coastal landscapes are formed.

The formation of headlands and bays

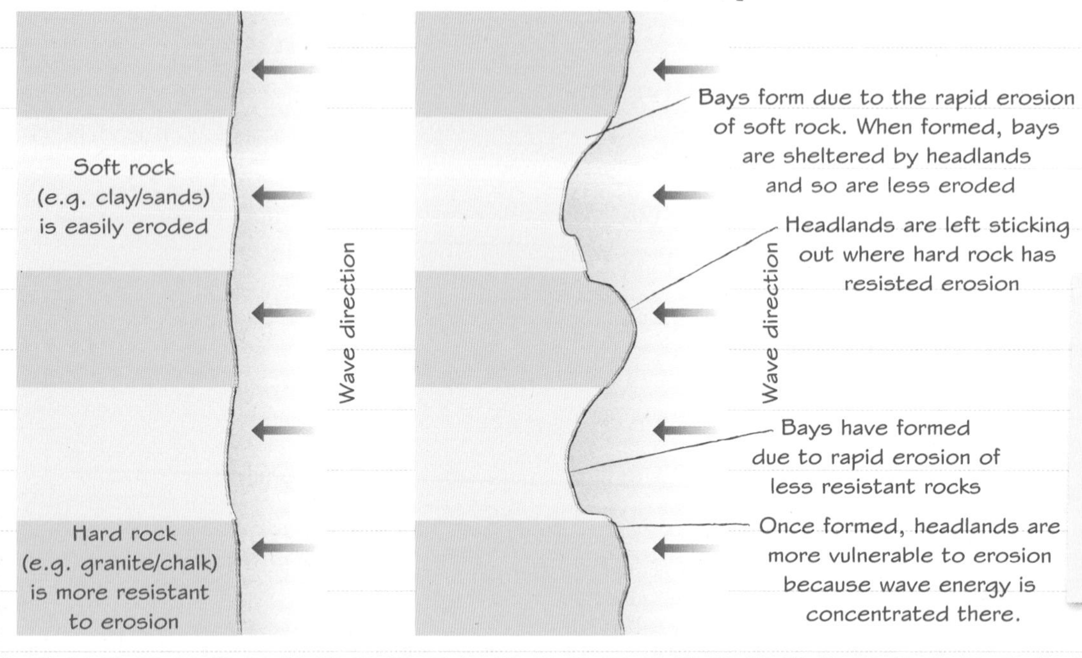

'Differential rates of erosion' is the technical term for when rocks of differing resistance are eroded at different rates.

Hard rock coastal landforms created by erosion

Caves, arches and stacks

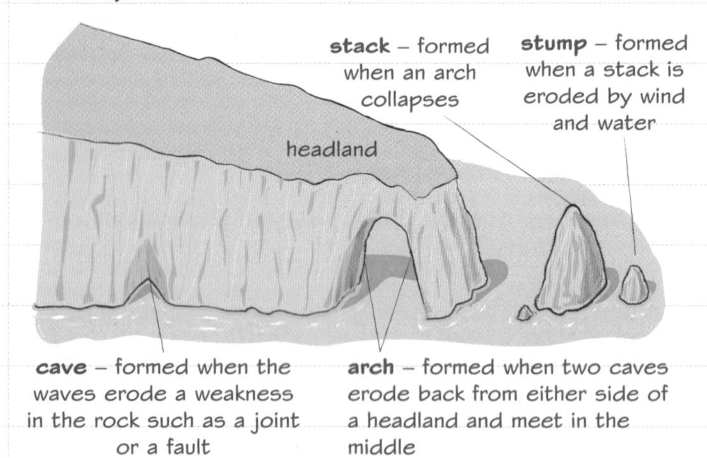

Wave-cut platforms

The erosion of cliffs can create wave-cut platforms – areas of flat rock at the base of the cliff.

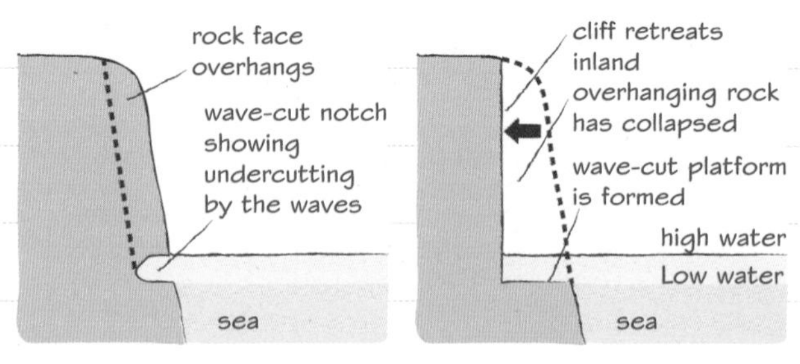

Worked example

Explain how geological structure can influence the erosion of a coastal headland. **(4 marks)**

Headlands are sections of resistant rock which jut out in the sea. Erosion by the sea will happen faster where there are gaps, cracks, joints, faults or other weaknesses in the rock. Caves are formed as erosion happens more quickly at this weaker section of the rock. If the weakness goes right through the rock, an arch may form as caves on either side of the headland join up.

Now try this

You could use diagrams to help you answer questions such as this one.

Explain how a wave-cut platform is formed. **(4 marks)**

Waves and climate

Destructive waves are the main cause of coastal erosion, but climate also plays an important role.

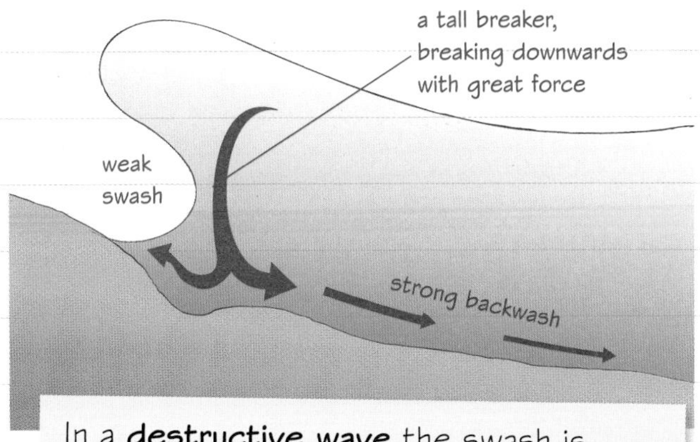

In a **destructive wave** the swash is weak and the backwash is strong, which means material is dragged back down a beach into the sea.

How waves erode the coast

- **Hydraulic action:** the sheer weight and impact of water against the coastline, particularly during a storm, will erode the coast. Also waves compress air in cracks in the rock, forcing them apart and weakening the rock.
- **Abrasion:** breaking waves throw sand and pebbles (or boulders) against the coast during storms.
- **Attrition:** the rocks and pebbles carried by the waves rub together and break down into smaller pieces.
- **Solution:** chemical action by seawater on some rocks, especially limestone.

Impact of UK climate on coastal erosion

The four **seasons** have different impacts on coastal erosion. For example, cold temperatures in winter lead to **freeze–thaw** weathering in cliffs.

Prevailing winds in the UK are from the south-west, bringing warm, moist air from the Atlantic and frequent rainfall, leading to weathering and mass movement on the coast.

Storm frequency is high in many parts of the UK, so coasts are often subject to strong winds, leading to an increase in the eroding power of waves and also leading to heavy rainfall contributing to mass movement.

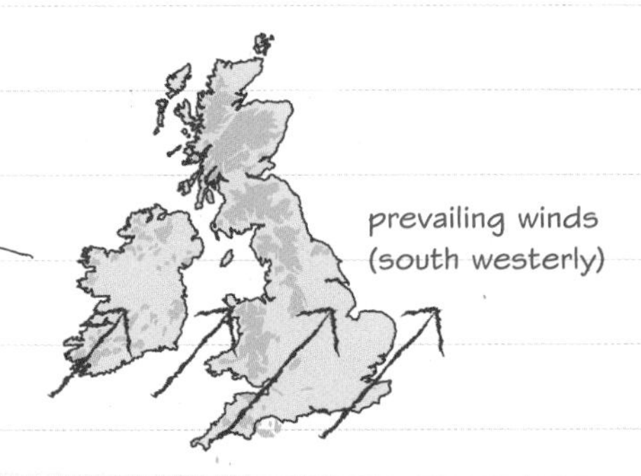

Worked example

This is a good answer because the points it makes are accurate and each explained in terms of contribution to coastal erosion.

Explain how the UK climate contributes to coastal erosion. **(4 marks)**

The UK's climate is temperate maritime, which means winters are mild and wet and summers are warm and wet. The prevailing winds from the south-west often bring rainfall to the country. The large amount of rainfall causes coastlines to erode through weathering and can also lead to mass movement and cliff collapse. Storm frequency is high, which brings heavy rainfall and strong winds that increase the erosional power of waves. The UK winter climate frequently sees temperatures dipping below freezing at night and then rising above 0°C in the daytime. When this is repeated many times, freeze–thaw weathering results, which adds to erosion.

Now try this

Describe **two** ways in which waves erode a coast. **(2 marks)**

Sub-aerial processes

Processes that impact on the land – such as weathering and mass movement – also contribute to coastal erosion.

Mechanical weathering

Freeze–thaw – most common in **cold** climates. When it freezes, water in cracks in the rock expands. Over time the crack widens and pieces of rock fall off. It is most effective when the temperature frequently rises above and falls below 0 °C.

Chemical weathering

This happens when the rock's mineral composition is changed.

- Granite contains feldspar which converts to soft clay minerals as a result of a chemical reaction with water.
- Limestone is dissolved by **carbonation**. Carbon dioxide in the atmosphere combines with rainwater to form carbonic acid, which changes calcium carbonate (limestone) into calcium bicarbonate. This is carried away by water in **solution**.

Biological weathering

This is caused by plants and animals and its action speeds up mechanical or chemical weathering. For example, tree roots widen gaps in rocks.

Mass movement

Mass movement is the downhill movement of material under the influence of gravity. The different types of mass movement depend on:

- the material involved
- the amount of water in the material
- the nature of the movement e.g. falls, slips or rotational slides.

Slumps happen when the rock (often clay) is saturated with water and slides down a curved slip plane.

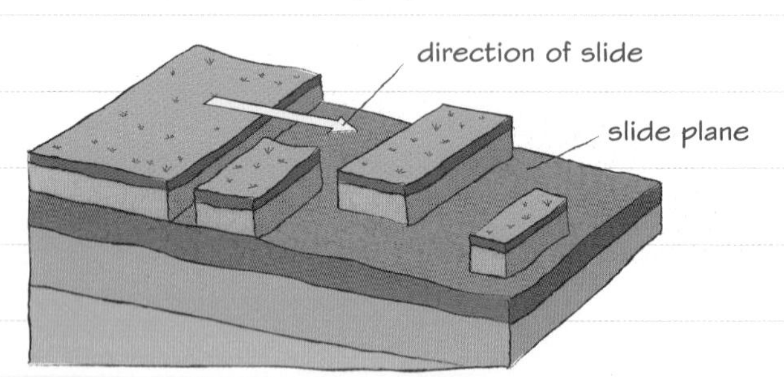

Sliding happens when loosened rocks and soil suddenly tumble down the slope. Blocks of material might all slide together.

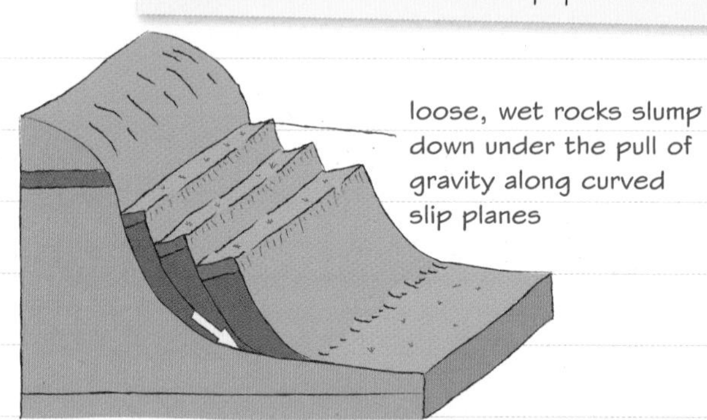

Coastal erosion leads to **coastal retreat**, when the coastline moves further inland. Another way to look at this question is to think about factors that increase the rate of erosion at the coast.

Worked example

Explain factors that lead to a fast rate of coastal retreat. **(4 marks)**

Coasts exposed to frequent storms will retreat faster than other areas because strong winds increase the eroding power of the sea and heavy rainfall will contribute to mass movement. Soft rock coastlines will retreat faster than hard rock coastlines as soft rock erodes quickly. Cliffs where the rocks have a large number of joints and faults will also erode more quickly than cliffs with fewer joints and faults.

This answer gives short but good, clear explanations of three factors.

Now try this

Draw a diagram to show the stages of freeze–thaw weathering. **(3 marks)**

Transportation and deposition

Waves transport eroded material along the coast and deposit it when they lose the energy to carry it further.

Longshore drift

1. Waves approach the coast at an angle.
2. **Swash** pushes sand and gravel up the beach at the same angle.
3. **Backwash** carries sand and gravel back down the beach at 90° to the coastline under the force of gravity.
4. Sand and gravel move along the beach in a zigzag fashion.
5. Sand is lighter than gravel so moves further up the beach.

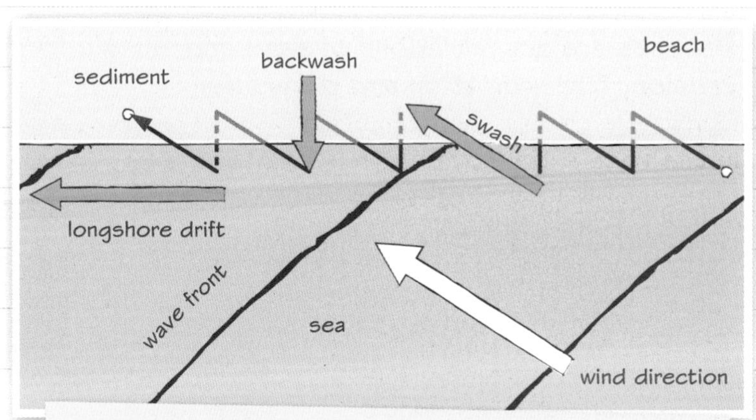

Adapted diagram courtesy of Barcelona Field Studies Centre, www.geographyfieldwork.com

Transportation

Waves **transport** material by:

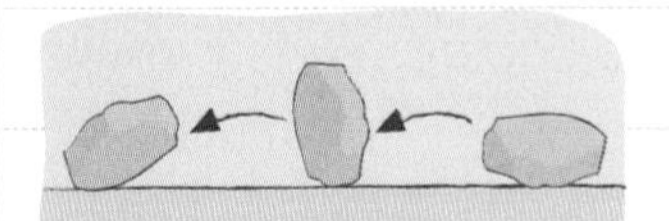

Traction – large boulders are rolled along the sea bed by waves

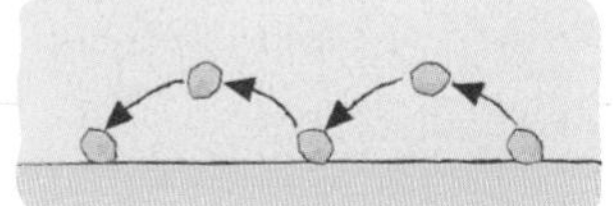

Saltation – smaller stones are bounced along the sea bed

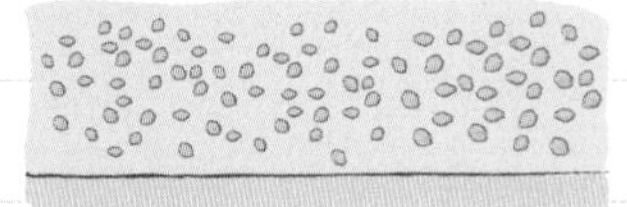

Suspension – sand and small particles are carried along in the flow

Solution – some minerals are dissolved in seawater and carried along in the flow

Deposition

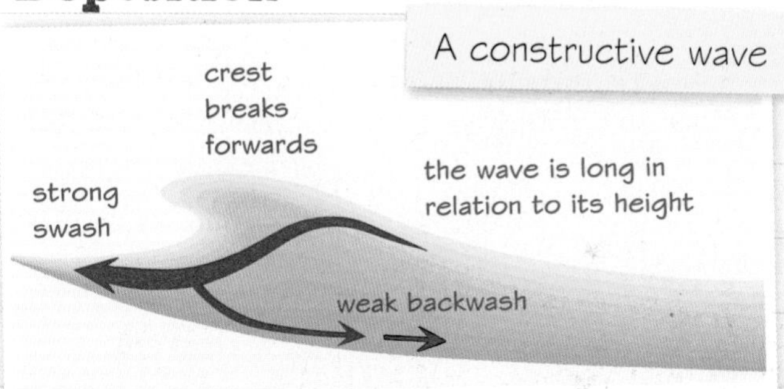

The load carried by waves is deposited by constructive waves. Different factors influence deposition, for example:

- sheltered spots (e.g. bays)
- calm conditions
- gentle gradient offshore causing friction.

All reduce the wave's energy.

Worked example

Describe the differences between a constructive wave and a destructive wave. **(4 marks)**

Constructive waves deposit material on beaches because they break gently on the beach and their strong swash carries material up the beach, while their weak backwash does not erode the material already on the beach. Destructive waves erode material from beaches because the backwash of these waves is much stronger than their swash and this drags material back down the beach into the sea.

This answer could have used annotated diagrams to show the main features of the waves.

Now try this

Explain the process of longshore drift. **(4 marks)**

Landscapes of deposition

Landscapes resulting from deposition include beaches, spits and bars.

Beaches

Beaches are accumulations of sand and shingle formed by deposition and shaped by erosion, transportation and deposition.

Beaches can be straight or curved. Curved beaches are formed by waves refracting, or bending, as they enter a bay.

Beaches can be sandy or pebbly (shingle). Shingle beaches are usually found where cliffs are being eroded and where waves are powerful. Ridges in a beach parallel to the sea are called berms and the one highest up the beach shows where the highest tide reaches.

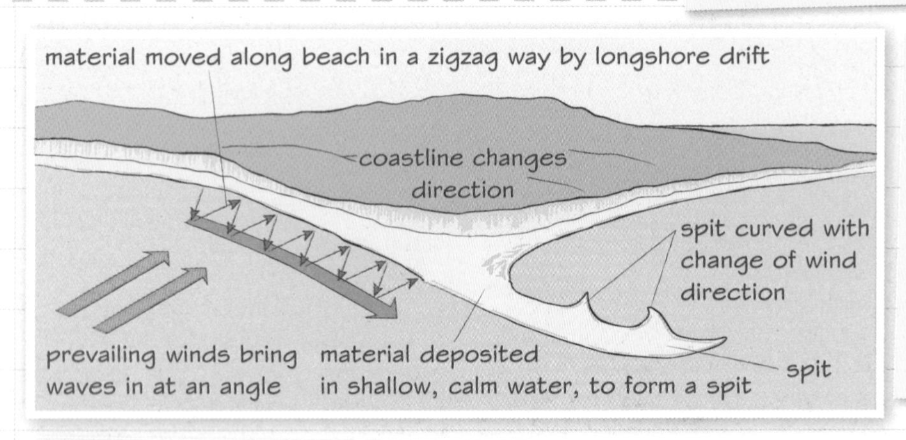

Spits are narrow projections of sand or shingle that are attached to the land at one end. They extend across a bay or estuary or where the coastline changes direction. They are formed by longshore drift powered by a strong prevailing wind.

Bars form in the same way as spits, with longshore drift depositing material away from the coast until a long ridge is built up. However, bars grow right across the bay, cutting off the water to form a lagoon.

Worked example

Identify the coastal landform shown on the map.

☐ **A** bar
☐ **B** beach
☐ **C** cave
☒ **D** spit

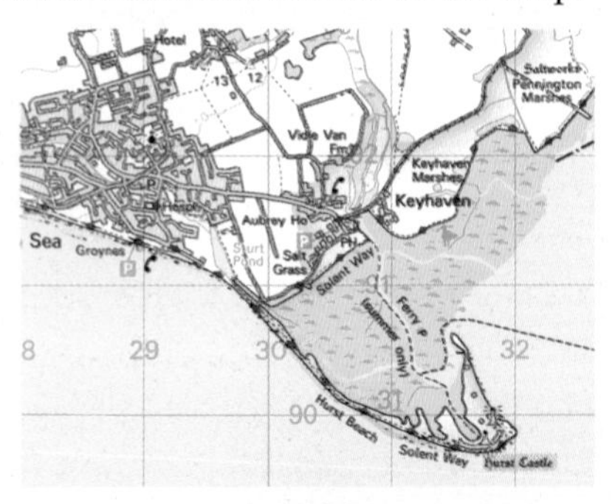

You need to be able to identify coastal landforms of deposition on OS maps like this one.

Now try this

Briefly describe how spits are formed.

(2 marks)

Human impact on coasts

Human activities can have direct and indirect, positive and negative, effects on coastal landscapes.

Development

- 👎 The weight of buildings increases cliff vulnerability
- 👎 Changes in drainage increase saturation
- 👍 Raises interest in protecting coastal landscapes

Industry

- 👎 Can cause/increase air, soil, water and noise pollution
- 👎 Can destroy natural habitats for birds, animals and sea life
- 👍 Brings wealth and jobs to an area

Agriculture

- 👎 Increased soil erosion
- 👎 Increased sedimentation
- 👍 Wildlife habitats may be created and preserved

Coastal management

- 👎 Can increase erosion further along the coastline
- 👍 Helps reduce risk of coastal flooding
- 👍 Some salt marshes, sand dunes, sand bars and spits are preserved and protected

Tourism

Coasts attract tourists for relaxing on beaches, swimming and water sports as well as enjoying the beautiful landscape. The effects can be diverse.

- 👎 Increased development for hotels and campsites impacts on natural processes – for example, increasing/decreasing coastal erosion, transportation and deposition and mass movement
- 👎 Increased pollution – for example, littering, noise, traffic fumes
- 👍 Increased revenue benefits people living there
- 👍 Increased desire to protect and preserve landscape so tourism continues

Worked example

Describe **two** effects of human activities on coastal landscapes. **(2 marks)**

Human development of coastlines by building houses and other buildings on the coast increases the risk of mass movement and cliff collapse as the weight of the buildings puts increased pressure on the cliffs and adds to run-off, which leads to soil saturation.

Humans can also decrease coastal erosion and transportation by building sea defences, which protect the coastline from the sea, reducing the risk of flooding and coastal retreat.

When a questions ask for a specific number of things (in this case 'two'), show the separate points, as in this example.

Now try this

Explain **one** way in which agriculture has affected coastal landscapes. **(2 marks)**

Holderness coast

Located example

You need to know how the interaction of physical and human processes is causing change on one named coastal landscape, including the significance of its location. We've used Holderness; you should use the example you did in class.

Holderness coast in East Yorkshire is one of the most vulnerable coastlines in Europe.

Significance of location

- Rock type is soft boulder clay, which is easily eroded and prone to slumping when saturated.
- Exposed to strong waves from the North Sea.
- Harder chalk rocks at Flamborough Head.

Physical processes at work

- Coastal erosion – a combination of strong waves (especially during storms) and rock type ensure the coast is eroded rapidly.
- The mean rate of erosion is around 2 m/year: one of the fastest-eroding coastlines in Europe.
- Mass movement – clay frequently slumps from the cliffs after rainfall.
- Transportation – strong waves move the eroded material away from the coastline.

Human processes at work

- Hard engineering on some parts of this coast – for example, rock armour and groynes at Mappleton have protected some areas from erosion and cliff collapse.
- Hard engineering in some places has prevented transportation, making erosion worse in other places.

Changes caused

Some parts of this coastline are retreating at a rate of nearly 2 m/year. Farmland, property and settlements have been lost to the sea, changing the landscape permanently.

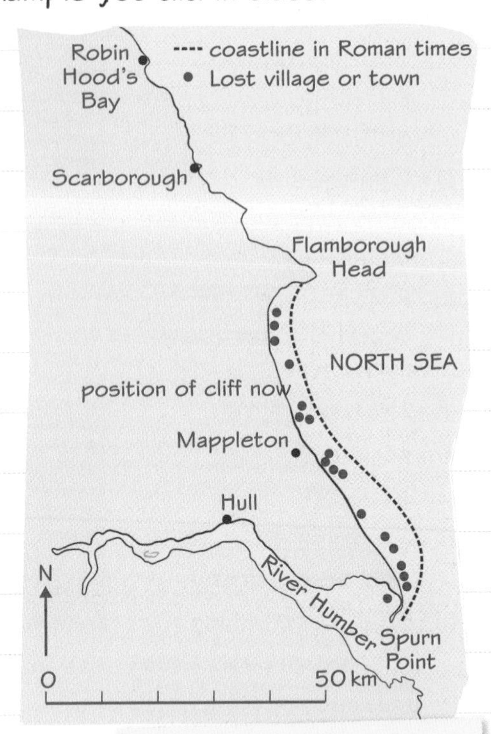

The Holderness coast

Sea defences at Mappleton on the Holderness coast

Worked example

Using one named example, describe **two** ways in which human processes have affected physical processes in a coastal landscape. **(3 marks)**

On the Holderness coast in East Yorkshire, hard engineering methods including rock armour and groynes have slowed down the rate of coastal erosion and cliff collapse, which would otherwise have been fairly rapid. However, the rock armour and groynes in some places, such as Mappleton, have prevented the transportation of material along the coast, leaving these other areas more exposed to erosion by the sea than they would have been and therefore speeding up the rate of coastal retreat.

Make sure you name your location in your answer.

Now try this

Explain how the location and geology of the Holderness coast contributes to its rapid rates of coastal erosion. **(4 marks)**

Coastal flooding

There is increased risk of coastal flooding in the UK that is mainly due to climate change.

Climate change

- As atmospheric temperature rises, it is likely that storm frequency and strength will increase. This can increase the height and strength of waves reaching the coast (especially when combined with high tides). An increase in heavy rainfall and wind will also increase weathering and mass movement on the coast.
- As sea temperature increases, the water expands so sea levels rise. The melting of ice on land also adds to the water in the ocean. Rising sea levels put low-lying coastal land at increased risk of flooding.

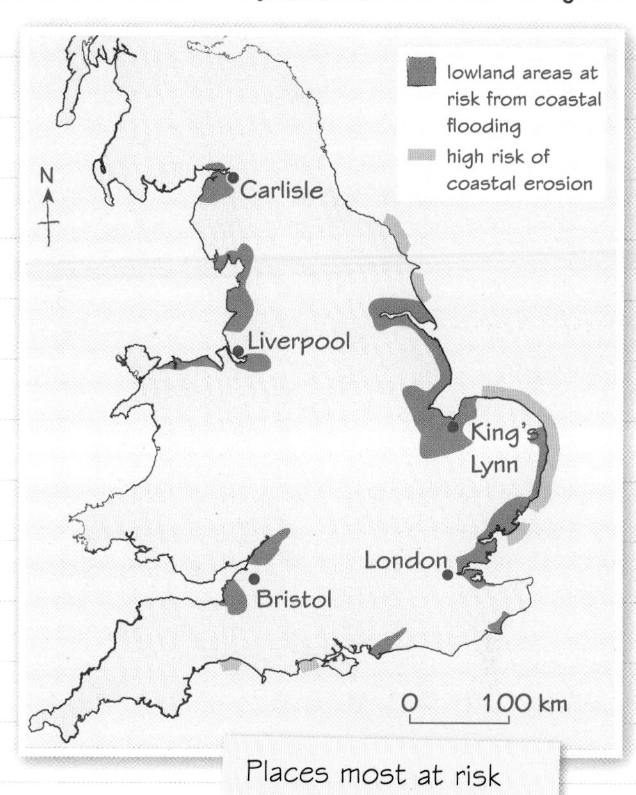

Places most at risk from coastal flooding in England and Wales.

The effects of climate change on the coastal environment

- Erosion may increase, so some beaches may disappear.
- Depositional features such as spits and bars may be submerged or destroyed.
- Natural ecosystems (e.g. the Essex marshes) and habitats may be destroyed.
- Erosion may increase, adding to coastal retreat and the risk of cliff collapse.

Worked example

Explain how climate change may affect coasts in the future. **(4 marks)**

Sea levels are predicted to rise because of climate change. This will mean that low-lying coastal areas, such as in much of East Anglia, are at increased risk of flooding or may even completely disappear into the sea.

There is likely to be more frequent and stronger storms in the UK. This will mean that coastlines are eroded faster and there may be more instances of coastal flooding, which may damage people's properties and destroy depositional features such as beaches.

This answer includes an example of an area that's at risk of increased flooding. It's always a good idea to include examples in your answers where you can.

Impacts of increased risk of flooding on people

- Flooding associated with storm surges can put people at risk of injury and death.
- Psychological impacts of losing or potentially losing homes and livelihoods.
- Settlements either need to be moved or defended, both of which will be expensive.
- Coastal tourism may diminish in some areas if beaches or other landscapes are lost.
- Flooding of roads and damage to railways will make travel more difficult.
- Loss of agricultural land will affect food production and the economy.

Explain why climate change brings an increased risk of coastal flooding in the UK. **(4 marks)**

Coastal management

Managing coastal processes can be done in different ways. All have different costs and benefits.

Hard engineering

Sea wall

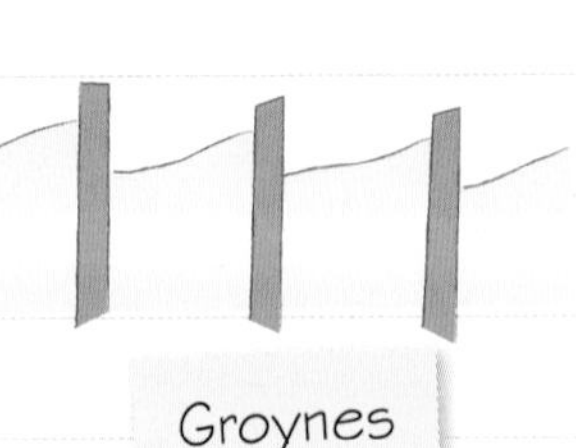

Groynes

- 👍 Protects cliffs and buildings
- 👎 Expensive

- 👍 Prevents sea removing sand
- 👎 Exposes other areas of coastline

Soft engineering

Beach replenishment

Slope stabilisation

- 👍 Sand reduces wave energy and maintains tourism
- 👎 Expensive

- 👍 Reduces slippage
- 👎 Foot of cliff still needs protection from the waves

Integrated Coastal Zone Management

Do nothing

- 👍 Cheaper than taking action
- 👎 Homes and land are lost

Hold the line

- 👍 Existing shoreline maintained
- 👎 Expensive

Strategic realignment

- 👍 People and activities move inland
- 👎 Unpopular with local residents

Worked example

Explain conflicting views on **one** method of coastal management. **(4 marks)**

Groynes are a hard engineering method of coastal management. They help prevent erosion of that bit of coastline and stop longshore drift transporting beach sediments along the coast. People who have homes and businesses that would be impacted by coastal erosion if the groynes were not built are likely to approve of the scheme. However, people who live further along the coast may disapprove as their homes and businesses will be negatively impacted. People who are not affected at all may disapprove of taxpayers' money being used to fund such schemes.

With questions like this, it is always good to begin answers by naming the method the answer will discuss.

Now try this

Outline one cost and one benefit of **one** hard engineering method of coastal management. **(4 marks)**

Investigating coasts: developing enquiry questions

Enquiry questions are the kind of questions that can be investigated by fieldwork in coastal environments. They give fieldwork a purpose. You will have put together enquiry questions for your fieldwork.

In your exam you will be asked questions about the fieldwork you did, and also some questions where you will need to apply what you learned on your fieldwork to new situations.

- An enquiry question often relates to a geographical theory: the sort of theory that can be tested through fieldwork.
- Key questions/hypotheses follow from the enquiry question, and they can be tested.

For example, an enquiry question could be:

- How do different methods of coastal management create benefits and conflicts?

A key question following on from this could be:

- Do people prefer hard engineering to soft engineering in managing coastal processes?

The enquiry process

There are six stages in the enquiry process and you will be asked questions on at least two of them in the exam.

Geographical examples and theories

You need to be able to identify the key geographical concepts that the investigation is based on.

For example, consider the enquiry question: **How and why do beach profiles vary in [name of coastal location]?**

To be able to evaluate different aspects of this investigation you would need to understand that beach profiles are affected by: wave type; wave frequency; wave direction/longshore drift; local geology; pebble size; beach management strategies (e.g. groynes).

Worked example

Explain **one** reason why Site 3 is not appropriate for comparing different coastal management strategies with Sites 1 and 2. **(2 marks)**

Site 3 is sheltered from waves approaching from the north-east and Sites 1 and 2 are sheltered from waves approaching from the south-west. The results from Site 3 will be different because of these factors rather than only because of different management approaches.

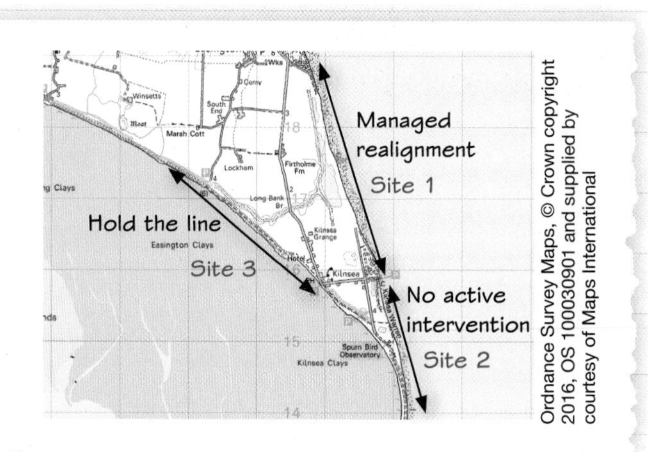

Now try this

Describe the location of your coastal management fieldwork. Explain why it was a good place to investigate coastal management and coastal processes. **(2 marks)**

Investigating coasts: techniques and methods

You will have used several different fieldwork techniques and methods in your investigation. You need to know what these techniques and methods are appropriate for, and what things to watch out for when using them, to avoid making errors in the field.

Worked examples

Explain **one** reason why the method you used to measure beach sediment was appropriate to the task. **(2 marks)**

Name of method used: Cailleux roundness index

It was appropriate because the beach was a pebble beach with a wide range of particle sizes and shapes. The Cailleux roundness index provided a consistent way of measuring pebble roundness.

For the exam you need to know about:

- one **quantitative** fieldwork method to measure how coastal management has affected beach morphology and sediment characteristics. Quantitative methods record data that can be measured as numbers.
- one **qualitative** fieldwork method to collect data on coastal management measures and their success. Qualitative methods record descriptive data, such as how people feel.
- you need to be able to say **why** the method you used was appropriate: there isn't just one correct method.

Explain **one** advantage of using secondary data to support your investigation. **(2 marks)**

We used a geology map. The geology map helped us choose good sites to study because it gave us information about where different types of rock were located along the coast.

For the exam you need to know about two secondary data sources:

- a geology map such as the Geology of Britain viewer
- one other source, which your teacher is likely to suggest for you.

Secondary data are data someone else has already collected. It will be useful for you to know about the different ways in which secondary data sources supported your investigation, and also about any particular advantages and disadvantages of your secondary data sources.

Explain **one** possible source of error when measuring beach sediment. **(2 marks)**

You need a way to make sure that you measure a representative sample of pebbles at each location along the beach profile. Using a quadrat helps you to avoid picking pebbles just because they look like they fit with your hypothesis.

You may be asked questions that require you to reflect on the methods you used and consider any ways in which problems could have occurred. For the exam, it will be useful to revise possible disadvantages of your coastal fieldwork methods.

Now try this

Study these two images. Which one shows a quantitative fieldwork method and which one shows a qualitative fieldwork method? **(2 marks)**

Measuring beach gradient with a clinometer

Interviewing residents and visitors on their views of coastal management measures

Investigating coasts: working with data

You need to know about ways to process and present fieldwork data, how to analyse it and how to make conclusions and summaries backed up by evidence from the data.

Worked example

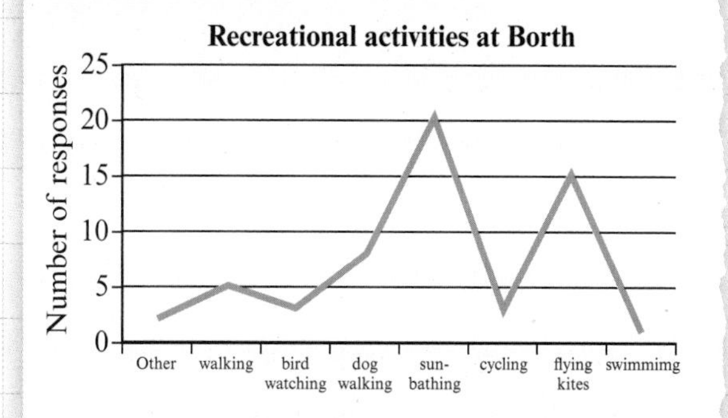

A student investigated what recreational activities people had come to Borth beach for, using a questionnaire survey. They presented their data using a line chart. Explain one weakness of using a line chart to present data from a questionnaire survey. **(2 marks)**

A line chart is appropriate for presenting continuous data. Survey data involving categories, like this fieldwork data, are discrete. A pie chart or bar graph would have been a more appropriate choice.

Data presentation disadvantages

This is a list of the top five disadvantages. You might be able to think of others, too.

1. Scattergraph: can only show relationships between two variables; inappropriate for more than two.
2. Pie charts: lots of small segments make the chart difficult to interpret.
3. Choropleth maps: hide variations within areas; give an impression of boundaries between areas instead of gradual transitions.
4. Triangular graphs: data must be in %.
5. Bar graphs: do not show relationships between categories.

Conclusions and summaries

The job of the conclusion is to use evidence from the investigation to answer the key question or hypothesis.

Analysing data

Key steps for successful data analysis are:

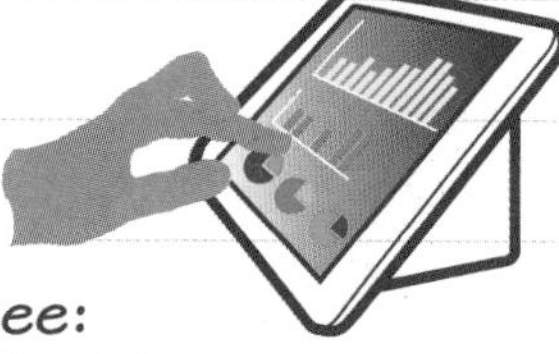

1. **Describe** what you see:
   - What are the overall patterns or main features?
   - Are any figures or features in groups?
   - What about anomalies or exceptions?
2. Use **evidence** – precise numbers or facts from the data – in your analysis.
3. Give **reasons** for the patterns you see in the data.
4. Link these reasons to **geographical** concepts/theories if you can.

Now try this

To measure beach morphology, students divided the beach into 10 segments by placing a ranging pole at each break of slope, and then they measured the length of each segment and its slope angle in degrees and minutes. Which one of the following would be an appropriate way of presenting their data?

☐ **A** beach transect diagram
☐ **B** Abney level
☐ **C** Power's scale of roundness
☐ **D** distribution diagram **(1 mark)**

Had a look ☐ Nearly there ☐ Nailed it! ☐

River systems

You need to know how river landscapes change between the different stages of a named UK river.

Characteristics of stages of the course of the Afon Nant Peris

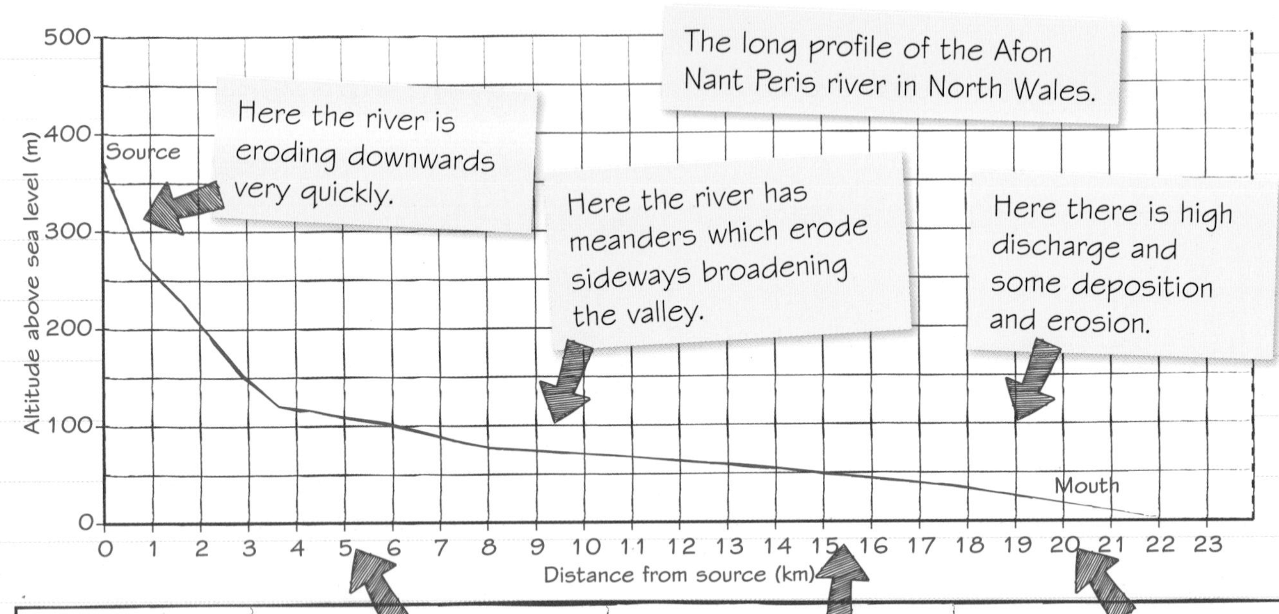

	Upper course	Middle course	Lower course
Gradient	steep	less steep	shallow gradient
Discharge	small	large	very large
Depth	shallow	deeper	deep
Channel shape	narrow, steep sides	flat, steep sides	flat floor, gently sloping sides
Velocity	quite fast	fast	very fast
Valley shape	steep sides	flat, steep sides	flat, gently sloping sides
Features	waterfalls, interlocking spurs	meanders, floodplain	meanders, floodplain, levees, ox-bow lakes

Worked example

Which **one** of the following is the correct description of how a river changes from source to mouth? **(1 mark)**

☐ **A** It gets steeper and narrower as it moves towards the sea.

☒ **B** It gets flatter and wider as it moves towards the sea.

☐ **C** It gets steeper and wider as it moves towards the sea.

☐ **D** It gets flatter and narrower as it moves towards the sea.

Now try this

Describe **one** change in gradient and one change in discharge along the course of a named UK river. **(3 marks)**

Start by underlining the most important words in the question.

Erosion, transportation and deposition

There are several processes at work in a river that interact to form river landforms. You need to know how the processes of erosion, transportation and deposition work.

The four main processes of **erosion**

Hydraulic action
The force of the water on the bed and banks of the river removes material.

Attrition
The load that is carried by the river bumps together and wears down into smaller, smoother pieces.

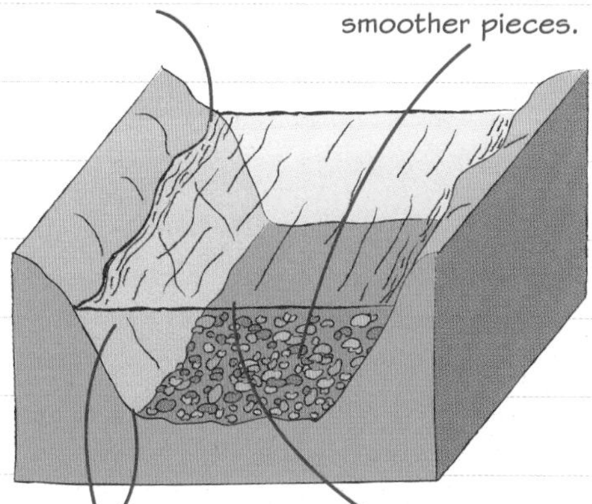

Abrasion
Material carried by the river rubs against the bed and banks and wears them away.

Solution
Some rock minerals dissolve in river water (e.g. calcium carbonate in limestone).

Worked example

Describe **one** type of river erosion. **(2 marks)**

Hydraulic action is one of the processes of erosion in a river. This is where the force of the water removes material from the river bed and banks, making the river deeper and wider over time.

Remember to include the name of the process in answers to questions like this. Use specialist terminology whenever you can.

Transportation

The four main types of **transportation.** Transportation is the way in which the river carries eroded material.

When the river loses energy (slows down) it may drop some of its load. This is called **deposition.**

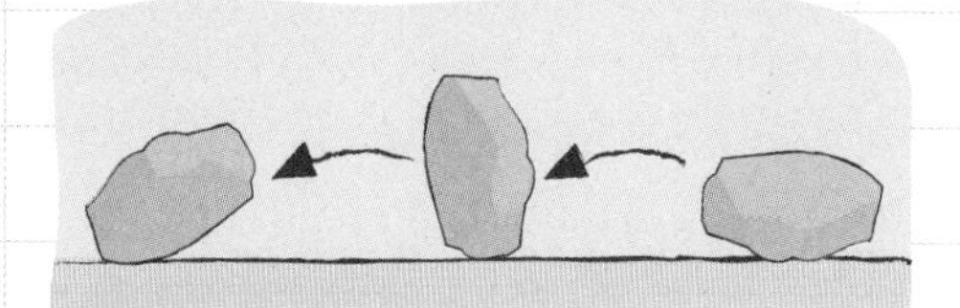

Traction: large boulders roll along the river bed

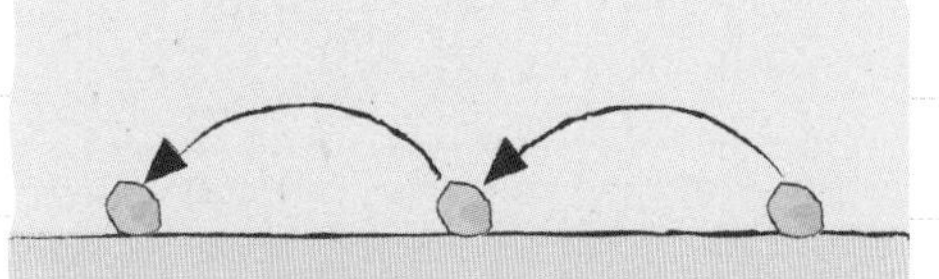

Saltation: smaller pebbles are bounced along the river bed, picked up and then dropped as the flow of the river changes

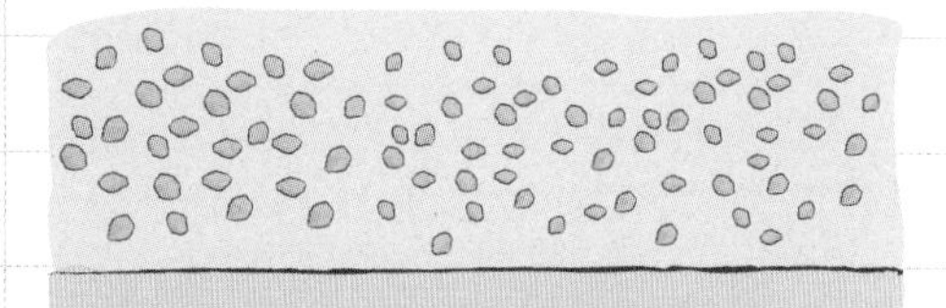

Suspension: finer sand and silt particles are carried along in the flow, giving the river a brown appearance

Solution: minerals from rocks such as limestone and chalk, are dissolved in the water and carried along in the flow, although they cannot be seen

Now try this

Describe **one** method of river transportation. **(2 marks)**

Had a look ☐ Nearly there ☐ Nailed it! ☐

Upper course features

In the upper course of a river, erosion is the main process at work. It creates various landforms including waterfalls and interlocking spurs.

Waterfalls

A waterfall is a steep drop in a river's course. The diagram explains how they are formed.

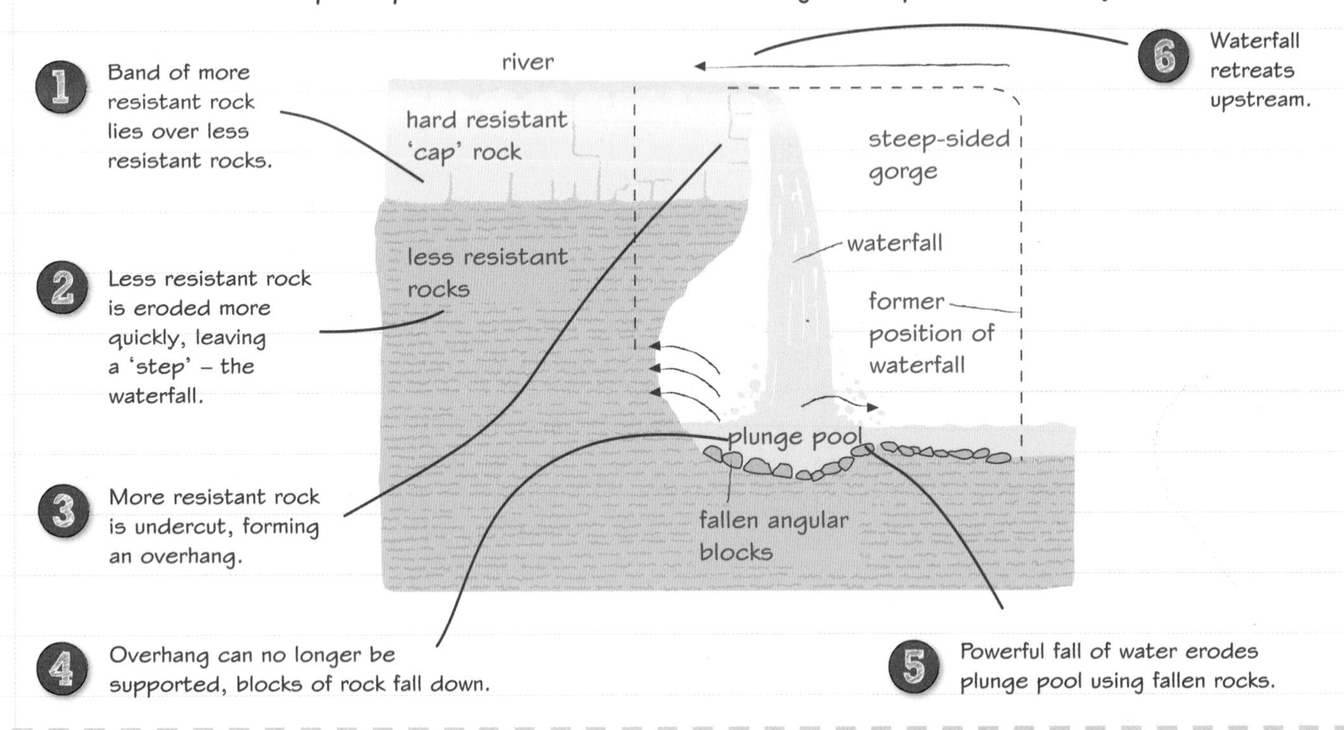

Interlocking spurs

The river at its source is small and has limited energy. It flows naturally from side to side, around ridges in the valley sides, called spurs. The spurs become interlocking with those on the other side of the valley.

Worked example

What **two** types of erosion are usually dominant in the formation of a waterfall plunge pool?

(2 marks)

Hydraulic action from the force of the waterfall hitting the bed of the plunge pool, and abrasion of the bed and sides of the pool caused by the rocks that have fallen down from the eroded hard cap of the waterfall.

Make sure you give the number of examples the question asks for, and name each process clearly using specialist terminology if you can.

Now try this

Briefly describe how interlocking spurs form. **(2 marks)**

Lower course features 1

Floodplains, meanders and ox-bow lakes are formed by erosion and deposition in the lower course of a river.

Meanders

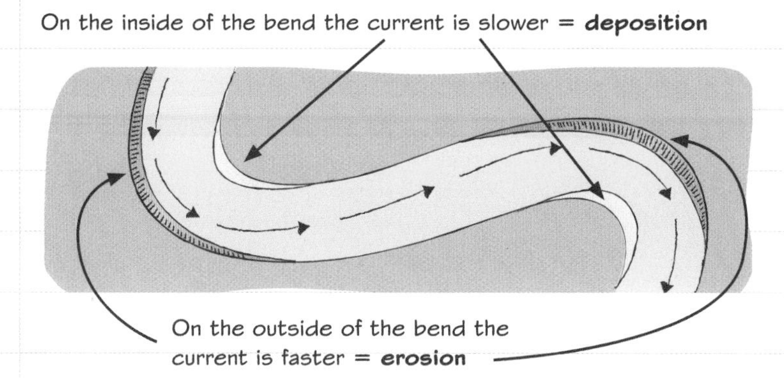

Meanders are bends in the river's course. In the lower course, the river uses up surplus energy by swinging one way and the other, causing lateral erosion on the outside of bends and deposition on the inside.

Ox-bow lakes

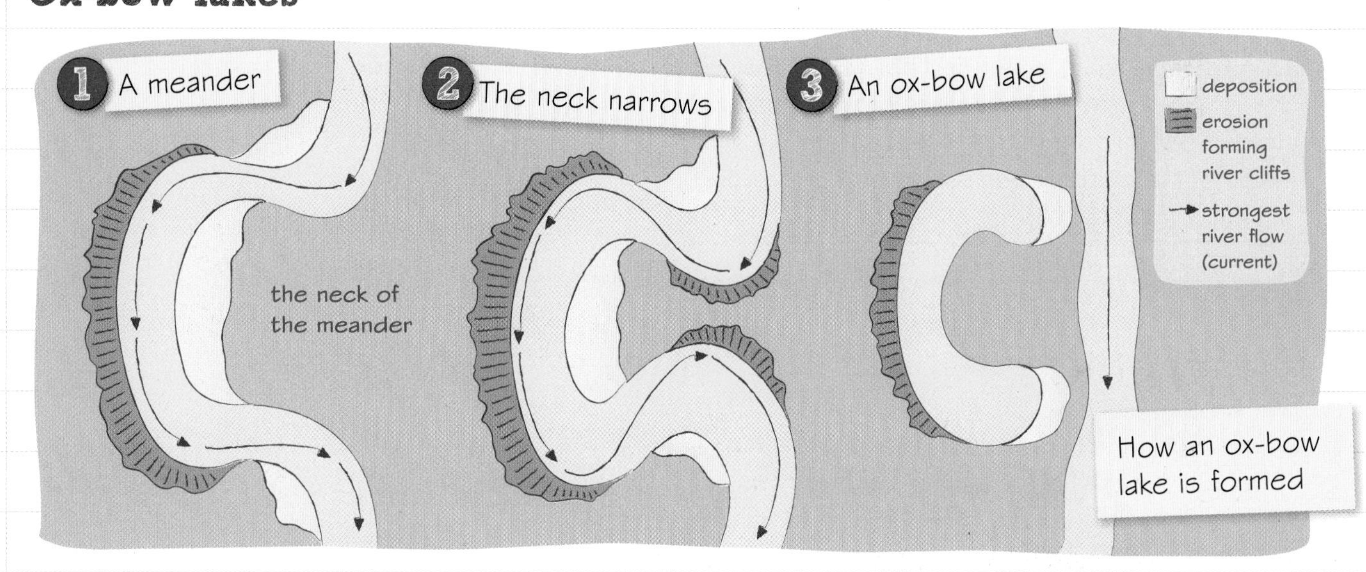

Worked example

Explain how erosion and deposition form floodplains. **(4 marks)**

As a river meanders in its lower course, lateral erosion erodes away the valley sides, making the valley flatter. The meanders migrate downstream. At the same time, deposition occurs as every time the river floods, fine particles of silt are deposited onto the valley sides, forming the floodplain over thousands of years.

Make it clear how you have answered all the parts of a question. In this answer, the student takes the two processes in the question – erosion and deposition – and tackles them separately.

Now try this

Briefly describe the main process affecting the lower course of a river. **(4 marks)**

Lower course features 2

Levées and deltas are formed by deposition in the lower course of a river.

Levées

As the river floods over its bank, the water slows down. The water can't carry the biggest and heaviest silt particles and they are dropped straight away on the bank forming floodplains.

Increased deposition on the river bed when the river is low gradually raises the river bed upwards.

After many floods, the deposits on the bank build up, forming levées.

Deltas

The speed of a river decreases as it approaches the sea and it deposits most of the material it has been carrying. Over time sediment builds up to create an almost flat area of new land, which is the delta. Because the river is now flowing very slowly over the almost flat gradient, its channel fills up with sediment and the river splits and spreads out into many different streams.

Worked example

Identify the river landform shown at grid reference 077198 on the map below. **(1 mark)**

☐ **A** waterfall

☒ **B** ox-bow lake

☐ **C** meander

☐ **D** levée

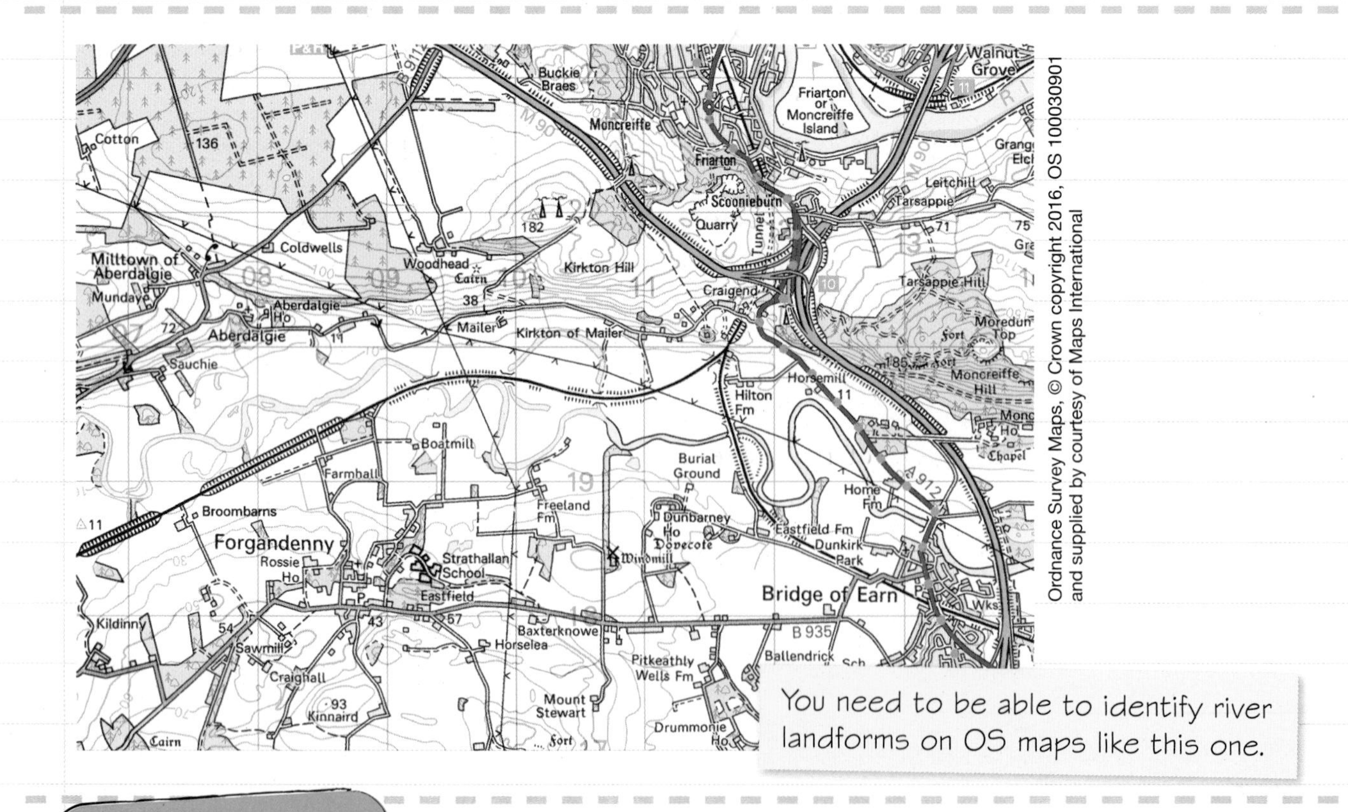

You need to be able to identify river landforms on OS maps like this one.

Now try this

Identify the river landform shown at grid reference 122198 on the map. **(1 mark)**

☐ **A** waterfall ☐ **B** ox-bow lake ☐ **C** delta ☐ **D** levée

Processes shaping rivers

Climate, geology and slope processes influence river landscapes and sediment load.

Impact of climate on river landscapes and sediment load

The **erosion rate** is greater where discharge and energy of river are greater – so rivers in wet climates erode more material than those in dry climates, impacting on the shape of the river valley and amount of sediment. (Discharge of river = volume of water flowing through river channel in a given time.)

The **transportation rate** is greater where energy of water is greater – so rivers in wet climates transport more material than those in dry climates.

Weathering of rocks is greater in some climates – for example, freeze–thaw weathering is most likely where temperatures frequently move from just above to just below freezing.

Amount of discharge is affected by climate. Wetter climates = greater discharge. Hotter climates = greater evaporation, so less discharge. The greater the discharge, the higher the transportation rate of the river.

Slope processes

Slope processes, also known as mass movement, affect the shape of river valleys and can increase sediment load in the river itself. There are two main types: **soil creep** and **slumping**.

Soil creep

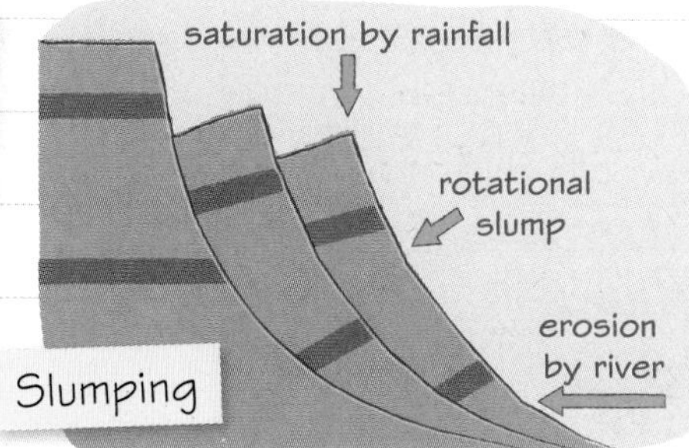

Slumping

Rivers flowing over resistant rock tend to have steep sides. Rivers flowing over less resistant rock tend to have gentle slopes.

1. Particles of soil slowly move down the sides of valleys under the influence of gravity.
2. Valley sides are eroded by the river making the sides steeper and increasing the downward movement of material. Heavy rainfall can trigger this movement.

Worked example

Explain how geology influences river landforms and sediment load. **(4 marks)**

River landforms are affected by the geology of an area. Some rocks, such as clay, are easily eroded by the river, whereas harder rocks, such as granite, are less easily eroded. This means that a river will erode its banks and bed, and transport and deposit more material where the rock is soft. This will mean the sediment load (the solid and dissolved material carried by the river) is greater than in an area of harder rock. Where there is a mixture of hard and soft rock, the river will carve a path through the less resistant rock first, creating features such as interlocking spurs. Waterfalls form where bands of hard rock overlie soft rock.

Note how the answer gives one point for sediment load and one point for river landforms and supports each point by explaining the influence of geology.

Now try this

Describe **two** slope processes that influence river landscapes. **(3 marks)**

Had a look ☐ Nearly there ☐ Nailed it! ☐

Storm hydrographs

Storm hydrographs are influenced by both human and physical factors.

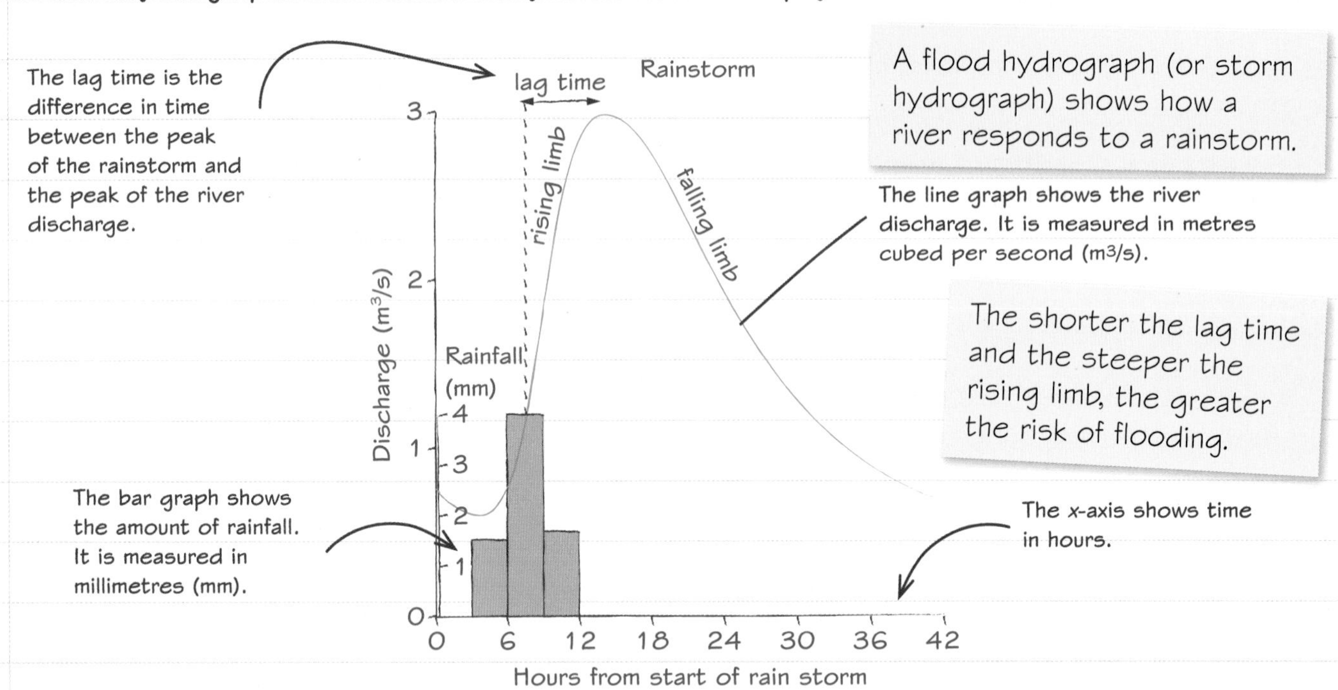

A flood hydrograph (or storm hydrograph) shows how a river responds to a rainstorm.

The shorter the lag time and the steeper the rising limb, the greater the risk of flooding.

Physical factors that make the lag time short, the rising and falling limbs steep, and discharge high on storm hydrographs include:

- Geology – more resistant rock will absorb less water than less resistant rock so run-off will be greater and faster.
- Soil type – as with geology, more impermeable soils (like clays) will absorb less water than permeable soils (like sandy soils) so run-off will be greater and faster. The amount of soil also has an impact.
- Vegetation – plants use water so run-off will be greater and faster on ground with less vegetation.
- Slope – steeper slopes cause faster surface run-off so more water reaches the river more quickly than on gentler slopes.
- Drainage basin shape – a wide basin with a lot of tributaries close together means water enters the river quicker making the rising limb steeper and lag time shorter.
- Antecedent conditions – when the ground is already saturated with water, further rain flows as run-off straight into the river.

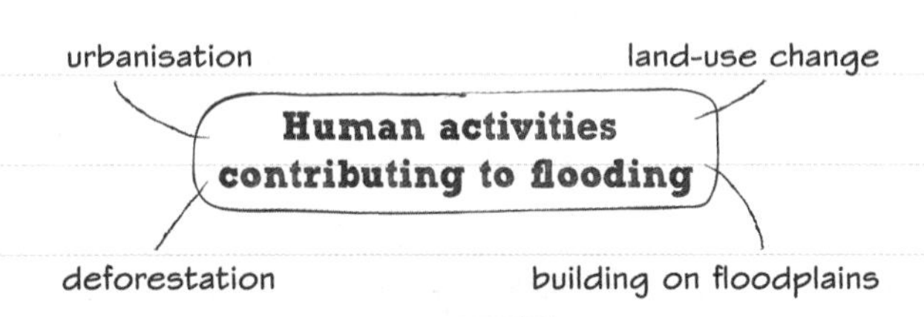

Worked example

Explain how human activities can alter storm hydrographs. **(4 marks)**

Deforestation and changes in land use reduce the amount of vegetation. Trees and other types of vegetation collect, store and use water. This reduces the lag time and makes the rising and falling limbs of the storm hydrograph steeper as water reaches the river faster than before.

Urbanisation means more impermeable surfaces replacing land that previously helped soak up water, causing a short lag time and steep rising limb.

This is a good answer that identifies two ways in which human activities can alter storm hydrographs, and explains each one.

Now try this

Explain how **two** physical factors affect storm hydrographs. **(4 marks)**

River flooding

You need to know how physical and human processes interact to cause flooding on one named river, including the significance of its location.

Tewkesbury in Gloucestershire frequently experiences river flooding from the River Severn.

Physical causes of flooding

Before the Severn reaches Tewkesbury it travels through many mountains and hills – steep slopes cause faster surface run-off.

The west of the UK frequently experiences heavy rainfall. Some believe that storms are increasing in severity and frequency due to climate change.

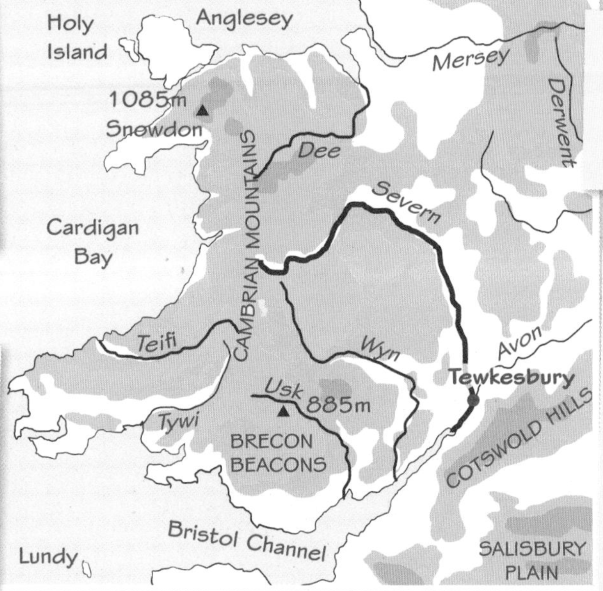

Two of the largest rivers in the UK – the Severn and the Avon – meet in Tewkesbury.

Snowmelt also contributes to the discharge of the Severn, especially in spring. The drainage basin of the Severn has lots of tributaries flowing into the river before it reaches Tewkesbury.

Human processes

- Urbanisation has caused the town of Tewkesbury and other towns along the course of the Severn to grow, meaning there is less soil to absorb water.
- In recent years, new housing has been built on the floodplain, which is extremely vulnerable to flooding. The old part of the town floods far less frequently as it is built on higher ground.
- Field drains on the mountains of Wales improve farmland, but they quickly move water into the tributaries of the Severn, contributing to flooding downstream.

Worked example

Explain how physical and human processes are causing river flooding on one named river.

(4 marks)

The River Severn floods regularly in Tewkesbury. Physical causes of the flooding include the location of Tewkesbury, which is at the meeting point of the River Severn and River Avon. Tewkesbury is also located toward the bottom of the River Severn's very wide drainage basin, which contains lots of tributaries close together and steep slopes, which makes run-off into the river faster. Human processes such as building on the floodplain and increasing urbanisation make the flooding worse.

This is a good example because the physical and human processes are quite specifically related to the named river.

Now try this

Explain how human processes are contributing to river flooding on **one** named river. **(4 marks)**

Increasing flood risk

There are increasing risks from river flooding and this is expected to continue in the future. Flooding brings many threats to people and the environment.

Climate change

Due to the Earth getting warmer there will be an increase in more extreme weather events. These can increase the risk of river flooding in the UK due to:

- increasing frequency of storms – more periods of heavy, intense rainfall meaning more water flowing into the river and increasing antecedent conditions
- increasing periods of hot, dry weather – bakes the soil so when it does rain the water runs off the surface as it can't soak in
- increasing periods of extreme cold – freezes the soil so water runs off the surface as it can't soak in.

Worked example

Explain how land-use change can increase the risk of river flooding. **(4 marks)**

Land-use change can increase the risk of river flooding in several ways. Where land is changed from soil to artificial surfaces, such as paving over gardens to create parking spaces, or building houses and roads on previously green spaces, this increases the amount of impermeable surfaces. This means less water can soak into the ground and it also increases the speed and amount of run-off, so more water reaches the river faster. Deforestation is another type of land-use change. This reduces the amount of interception so rain reaches the ground faster and surface run-off increases. Building on floodplains removes one way that rivers can flood without causing damage, and increases the risk of flooding as the amount of impermeable surfaces increases.

Focusing on the key parts of the question is really important. Here the student correctly zooms in on land-use change increasing the risk of river flooding, rather than examining increased risks of river flooding generally.

What are the threats of flooding to people and the environment?

- Damage to homes
- Damage to farmland and crops causing problems with food supply
- Death and injury to people
- Damage to roads, bridges and railways causing problems with transport
- Damage to businesses through destruction of goods and property, and transport and communication delays
- Damage to freshwater and electricity supplies

- Death of, and injury to, animals and fish
- Damage to wildlife habitats
- Reduction in birds' and animals' food supply
- Pollution damaging the land if sewage and chemicals are in the floodwater
- Damage to and destruction of plants and trees
- Stress to animals that have to be rescued from flooded areas

Now try this

Identify **two** threats of flooding to people. **(2 marks)**

Managing flood risk

Hard and soft engineering are used to help manage the risk of river flooding, but each method has costs and benefits.

Hard engineering methods

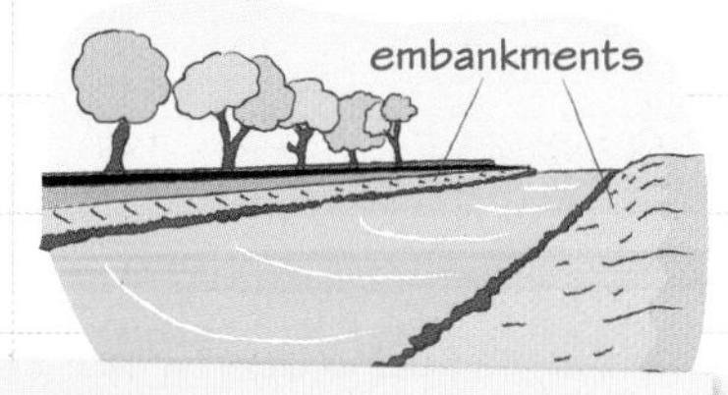

Embankments and levées

- 👍 Use natural materials so blend in with surroundings
- 👎 May burst, causing widespread flooding

Flood walls

- 👍 Require minimal maintenance
- 👎 Block the view of the river

Dams and reservoirs

- 👍 Able to regulate and control the flow of water
- 👎 Very expensive to build

Flood barriers

- 👍 Can be moved to where needed and quickly erected
- 👎 Don't provide long-lasting protection

Soft engineering methods

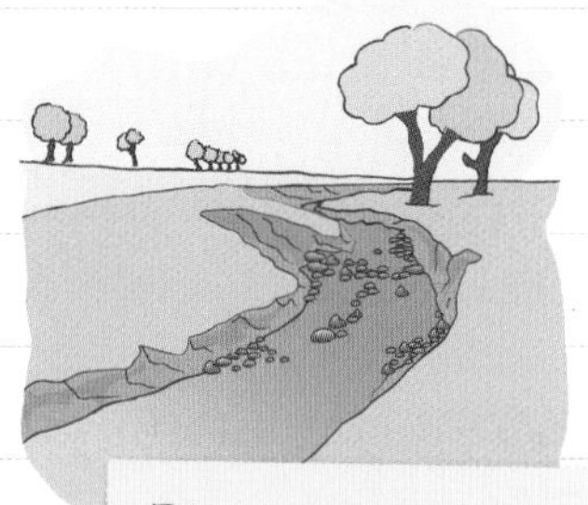

River restoration

- 👍 Can reduce flooding downstream
- 👎 People living there may not want land use to change

Washlands

- 👎 Restricts economic development

Floodplain retention

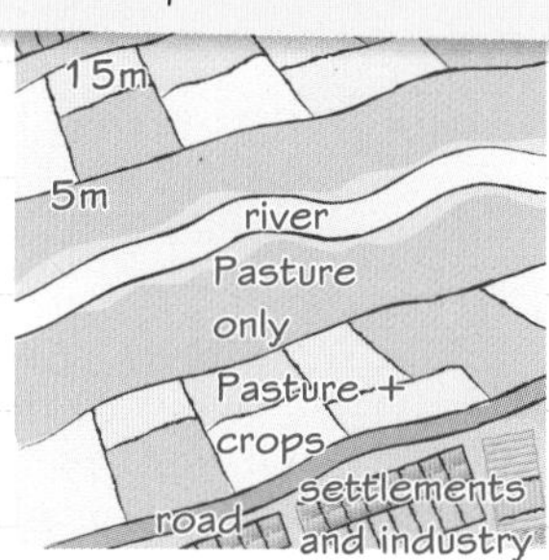

- 👍 Provides somewhere for floodwater to go
- 👎 Restricts economic development
- 👍 Attractive and provides space for leisure and recreation

Plant trees (afforestation)

- 👍 Increased infiltration
- 👎 Not suitable for all locations

Worked example

Explain one cost and one benefit of **one** hard engineering approach to managing river flooding. **(4 marks)**

A benefit of embankments is that they are constructed from natural materials so can be made to blend in with the environment and look attractive. A cost is that they require regular inspection and maintenance, which can be expensive.

Now try this

Explain **one** cost and **one** benefit of a soft engineering approach to managing river flooding. **(4 marks)**

Investigating rivers: developing enquiry questions

Enquiry questions are the kind of questions that can be investigated by fieldwork in river environments. They give fieldwork a purpose. You will have put together enquiry questions for your fieldwork.

In your exam you will be asked questions about the fieldwork you did, and also some questions where you will need to apply what you learned on your fieldwork to new situations.

The enquiry process

There are six stages in the enquiry process and you will be asked questions on at least two of them in the exam.

- ✓ An enquiry question often relates to a geographical theory: the sort of theory that can be tested through fieldwork.
- ✓ Key questions/hypotheses follow from the enquiry question, and they can be tested.

For example, an enquiry question could be:

- ✓ How do the river valley and channel characteristics vary along the River [name]?

A key question following on from this could be:

- ✓ Do the width and depth of the river increase as the river flows downstream?

Geographical examples and theories

You need to be able to identify the key geographical concepts that the investigation is based on.

For example, for the enquiry question: **How do the river valley and channel characteristics vary along the River [name]?**

Geographers would use Bradshaw's model. This is a theoretical model that sets out the changes we expect to see from source to river mouth, for example:

- ✓ discharge increases downstream
- ✓ channel depth increases
- ✓ load particle size decreases
- ✓ slope angle decreases.

Worked example

Study this figure, which shows five sites selected by students along the Hawkcombe Stream.

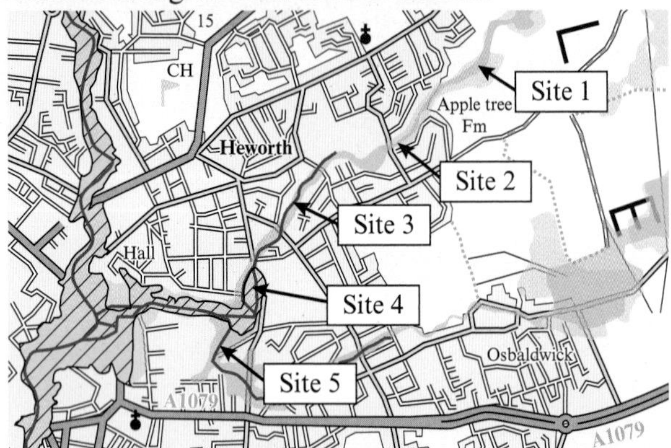

Explain one reason why Site 5 is not appropriate for comparing the impact of changing river discharge and drainage basin characteristics on flood risk with Sites 1–4. **(2 marks)**

Site 5 is on a different tributary of the stream, so its discharge and characteristics would not be comparable to sites 1–4.

Answering this question well depends on having good geographical understanding of river processes.

Now try this

Describe the location of your river fieldwork. Explain why it was a good place to investigate how and why drainage basin and channel characteristics influence flood risk for people and property. **(4 marks)**

Investigating rivers: techniques and methods

You will have used several different fieldwork techniques and methods in your investigation. You need to know what these techniques and methods are appropriate for, and what things to watch out for when using them, to avoid making errors in the field.

Worked examples

1 Explain **one** reason why the method you used to measure stream depth was appropriate for the task. **(2 marks)**

Name of method used: A metre ruler and tape measure

It was appropriate because the stream we worked on was not deep (it was under 30 cm), allowing for accurate measurement with a metre ruler.

The specification says that for the exam you need to know about:

- one **quantitative** fieldwork method to measure changes in river channel characteristics. Quantitative methods record data that can be measured as numbers.
- one **qualitative** fieldwork method to collect data on factors that might influence flood risk. Qualitative methods record descriptive data such as how people feel.
- you need to be able to say why the method you used was appropriate for the task because there isn't just one method of making a measurement.

2 Explain **one** reason why using secondary sources of data supported your investigation. **(2 marks)**

We used a flood risk map. The flood risk map helped us choose good sites to study because it gave us information about where areas and property were judged to have an increased chance of flooding if the stream was to burst its banks.

For the exam you need to know about two secondary data sources:

- a flood risk map, such as an Environment Agency flood risk map
- one other source, which your teacher is likely to suggest for you.

Secondary data are data someone else has already collected. It will be useful for you to know about the different ways in which secondary data sources supported your investigation, and also about any particular advantages and disadvantages of your secondary data sources.

3 Explain **one** possible source of error when you measured stream depth. **(2 marks)**

I used a metre ruler. If the stream bed had been sandy, the ruler might have sunk into it and this could have introduced errors.

You may be asked questions that require you to reflect on the methods you used and consider any ways in which problems could have occurred. For the exam, it will be useful to revise possible disadvantages of your river fieldwork methods.

Now try this

Study these two images. Which one shows a quantitative fieldwork method and which one shows a qualitative fieldwork method? **(2 marks)**

Measuring water quality by sampling

Collecting views on river flooding by questionnaire

Had a look ☐ Nearly there ☐ Nailed it! ☐

Investigating rivers: working with data

You need to know about ways to process and present fieldwork data, how to analyse it and how to make conclusions and summaries backed up by evidence from the data.

Worked example

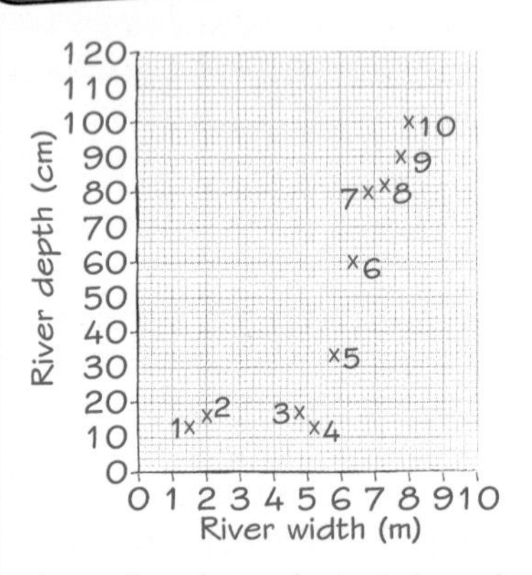

A student investigated the relationship between river depth and river width. They presented their data using a scattergraph. Explain **one** advantage of using a scattergraph to present these data. **(2 marks)**

An advantage of a scattergraph is that it can be used to identify a relationship between two variables. A line of best fit can be drawn to indicate whether the relationship is positive, negative or whether there is no relationship.

Advantages and disadvantages

The exam may ask questions about how the presentation of fieldwork data could be improved. Use your knowledge of what the advantages and disadvantages are of different types of data presentation.

Presentation disadvantages

This is a list of the top five disadvantages. You might be able to think of others, too.

1. Scattergraph: can only show relationships between two variables.
2. Pie charts: lots of small segments make the chart difficult to interpret.
3. Choropleth maps: hide variations within areas.
4. Triangular graphs: data must be in %.
5. Bar graphs: do not show relationships between categories.

Analysing data

Key steps for successful data analysis.

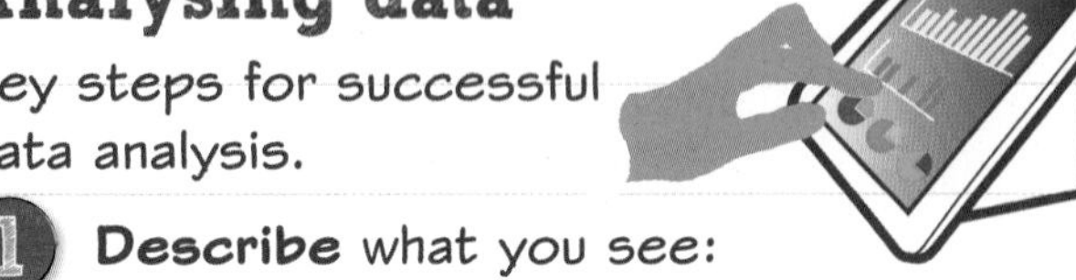

1. **Describe** what you see:
   - What are the overall patterns or main features?
   - Are any figures or features in groups?
   - What about anomalies or exceptions?
2. Use **evidence** – precise numbers or facts from the data – in your analysis.
3. Give **reasons** for the patterns you see in the data.
4. Link these reasons to **geographical** concepts/theories if you can.

Conclusions and summaries

The job of the conclusion is to use evidence from the investigation to answer the key question or hypothesis.

Now try this

The student used a dispersion graph to compare the size of pebbles from the river bedload at five different sites. Suggest **one** advantage of presenting the data this way. **(1 mark)**

Urban and rural UK

You need to know some key differences between urban core and rural places and how UK and EU government policies have attempted to reduce these differences.

UK population density

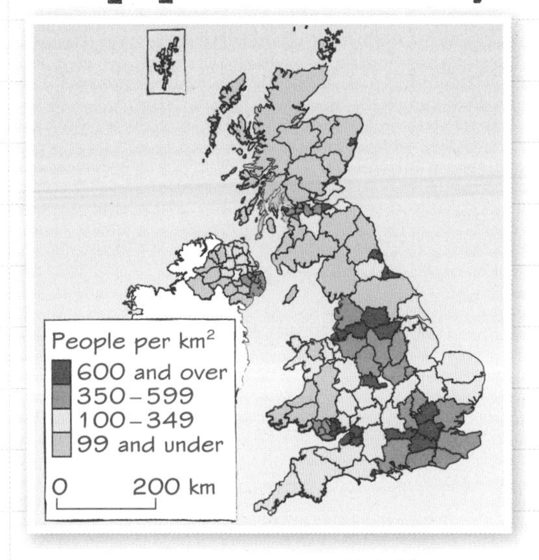

Definitions

Rural settlements – in the UK, settlements with fewer than 10 000 inhabitants (fewer than 3000 in Scotland)

Conurbation – when a city has expanded outwards and absorbed smaller settlements that used to be separate

Urban core – the central part of a conurbation: high population density

Urban fringe – the settlement areas around the edge of the urban core: lower density

Population density – the number of people per square kilometre

Worked example

Study the population structure diagram opposite, which shows the age structure of UK urban and rural areas in 2001 and 2011.

Explain **two** reasons why the rural population structure has changed between 2001 and 2011. **(4 marks)**

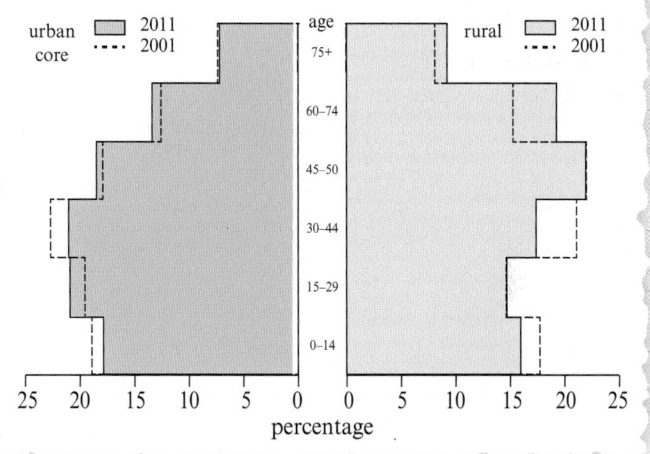

People who work in cities often move to rural areas when they retire because of rural area attractions: quieter, less congested, natural.

Younger people who grow up in rural areas tend to move to urban areas because of wider job opportunities, higher pay, better services.

Economic differences

- There are more people working in the primary sector in rural areas: agriculture, forestry and fishing.
- Many of the people who live in rural areas work in urban areas.
- Rural settlements have lower average wages than urban cores.

UK and EU policies

Policies to reduce economic differences between urban and rural places include:

- The EU's European Regional Development Fund – £2.6 billion (England). The EU invests in businesses in poorer regions to help them grow.
- Enterprise Zones – tax cuts to attract businesses to specific regions, plus superfast broadband.

Now try this

The average hourly pay in Wales in 2011 was £9.10 while in London it was £17.00.

Describe **two** ways in which the UK government aims to reduce regional differences like this within the UK.

(2 marks)

The UK and migration

You need to know about migration numbers, distribution and age structure, and ways that immigration has increased ethnic and cultural diversity.

International migration to the UK

Immigration has increased over the last 50 years.

- ✓ In 1961, 3 per cent of people living in the UK were born in another country.
- ✓ In 2015, 13 per cent of UK residents were born in another country.
- ✓ In 2015, 8 million people born outside the UK lived here.
- ✓ In 2015, UK **net immigration** (the difference between those immigrating and those emigrating) was estimated at 336000.

Immigration and age structure

Most immigrants are young and therefore more likely to have children. This influences UK age structure.

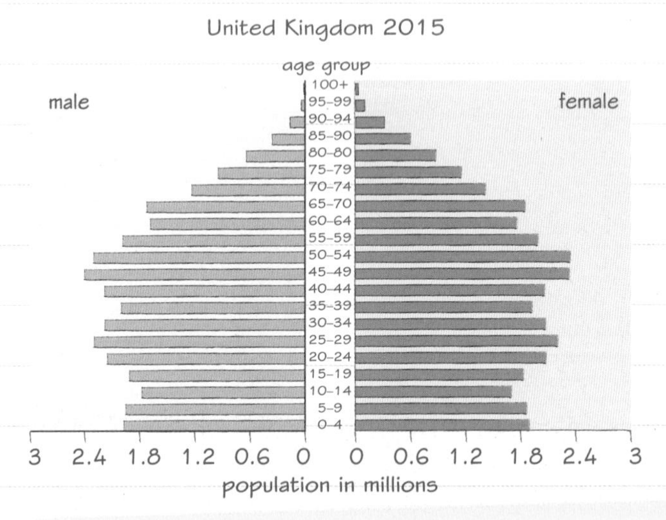

- In 2014, 25 per cent of people living in Inner London were aged 25 to 34 compared with 13 per cent in the rest of England.
- 27 per cent of births in the UK in 2014 were to mothers born outside the UK.

Immigration policy

- Current (2016) UK government policy is to reduce the level of net migration to 100000 people per year.
- However, while the UK is a member of the European Economic Area (EEA), the government cannot restrict the movement of EEA citizens to the UK.
- This means the UK's policy is to make immigration by non-EU people more difficult.

Immigrants come from many different EU and non-EU countries: the top five are India, Poland, Pakistan, Ireland and Germany.
London was the most popular destination for immigrants to the UK in 2012.
Northern Ireland was the only UK region with more emigrants than immigrants.

Worked example

Study the age structure diagram for the UK in 2015 above.

Describe **two** ways in which immigration influences age structure. **(2 marks)**

Immigrants are usually young – in their 20s. This means immigrants can increase the number of 20 to 30-year-olds in a population structure.

This age is when most people have families, so immigration also tends to increase the number of babies and young children in a population.

Two clear points that correctly describe two influences.

Now try this

Explain **one** way in which European policy on immigration influences cultural and ethnic diversity in the UK. **(2 marks)**

Economic changes

Some regions of the UK that depended on primary and secondary sectors have now become successful tertiary centres, while other regions have not. You need to know about the differences in contrasting UK regions, for example north-east and south-east England.

There has been a big growth in the importance of tertiary and quaternary industry in the UK in the last 50 years.

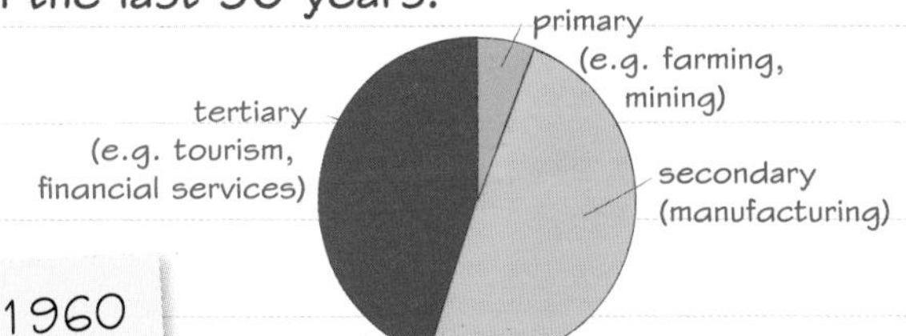

quarternary (e.g. IT services)
primary (e.g. farming, mining)
tertiary (e.g. tourism, financial services)
secondary (manufacturing)
2011

The decline of the coal industry

In 2015 the last deep coal mine in the UK was closed. Although coal is still an important fuel in the UK, it is much cheaper to import it from other countries than use coal mined in the UK.

Some regions in the UK, such as the North East and South Wales, had a long history of coal mining. Whole communities depended on coal mining for work.

Where pits have closed down, new service industries have grown up, such as warehousing. However, wages in these industries are much lower than the coal miners had been able to earn previously.

Miners at Kellingley Colliery in December 2015 as the pit – the last UK deep coal mine – was closed.

Worked example

Study the fact file about the London Docklands.

Explain **one** reason why London has been able to deal with the decline in secondary industry.

(2 marks)

London was already a global financial centre as well as an important manufacturing city. When secondary sector employment declined, London's population declined too. Government investment (£1.8 billion) to encourage the redevelopment of the London Docklands allowed the financial industry (tertiary sector) to expand and create new jobs.

Fact file: London Docklands

- In the 1930s, the London docks were the world's largest. Products came to London from all over the British Empire.
- Global trade moved to container ships: these ships were too big for London docks.
- Between 1951 and 1981, 100 000 jobs dropped to just 27 000 in the Docklands.
- Jobs in manufacturing declined by 80 per cent in London after 1960 as manufacturing moved out of London and then abroad.
- In 1981, the government invested £1.8 billion on regenerating the Docklands.
- There was high demand for new offices. 100 000 new jobs were created in financial services and business services.

This answer uses relevant detail (e.g. £1.8 billion) to enhance the reason given.

Now try this

Describe **two** reasons why the number of people employed in the primary sector has declined in the UK. **(2 marks)**

Had a look ☐ Nearly there ☐ Nailed it! ☐

Globalisation and investment

Globalisation has meant increased FDI in the UK, including investment by TNCs in the UK economy.

Definitions

Globalisation – the process by which trade and investment build more and more connections between countries

Foreign direct investment (FDI) – when people in one country invest in businesses in another country to the extent that they gain significant control over how those businesses are run

Transnational corporations (TNCs) – businesses run from one country that have control over enterprises in other countries

Worked example

Study the map below, which shows the level of FDI in European countries in 2012.

Identify which **one** out of the following countries from the map had the highest level of FDI in 2012. **(1 mark)**

☐ **A** UK
☐ **B** Denmark
☒ **C** Belgium
☐ **D** Germany

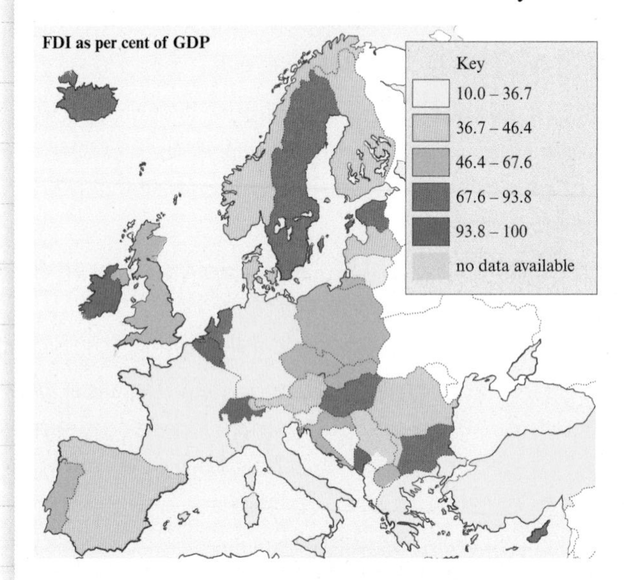

Why has FDI in the UK increased?

FDI in the UK increased by 88 per cent between 2005 and 2015. Why?

- Globalisation is strong in banking and finance because money can be moved electronically around the world. London is a global centre for finance.
- Trade deals with developed and emerging countries make imports and exports cheaper and easier and create UK jobs.
- The EU encourages **free trade** between member states and sets up good trade deals for the whole EU with other countries.
- UK governments have encouraged FDI by privatising industries and allowing foreign companies to buy them.

Role of TNCs

TNCs have advantages and disadvantages for the UK economy.

- Advantage 1: **investment**. Foreign companies have invested £1 trillion in the UK, creating thousands of jobs.
- Advantage 2: **innovation**. Foreign companies bring new technologies.
- Disadvantage 1: **security**. Economic problems abroad can mean production gets shut down in the UK: job losses.
- Disadvantage 2: **competition**. TNCs can outcompete UK companies due to the TNCs' massive economies of scale.

Now try this

TNCs are able to reduce the amount of tax they pay on their sales in countries like the UK by setting up headquarters in low-tax countries like Eire or Luxembourg. Explain why this is seen as a disadvantage of TNCs. **(4 marks)**

A UK city in context

Case study For this case study, you will have learned about how a major UK city is changing. We use Birmingham as an example of a UK city here, but you should revise the city that you did in class.

City in context

Make sure you know about why your case study city is important: in the UK; in its region; and globally. You should consider its **site**, **situation** and **connectivity**.

For example, Birmingham is:
- located in the centre of **England**, with fast motorway and rail connections
- the main city of the West Midlands **region**, a major centre of manufacturing
- home to people from all over the **world** – one third of residents are from minority ethnic backgrounds.

If you are doing **urban areas** for one of your geographical investigation options, you will have done fieldwork and research on an urban area too. See pages 77–79 of this book for more on r the urban geographical investigation option.

If you are doing **rural settlements** for your geographical investigation, **you still need to revise your case study of a major UK city**. Then you need to see pages 80–82 for more on the rural geographical investigation option.

City structure

How do the different parts of your city work – what is their function? Functions include: residential, industrial, tourism, retail, cultural or religious, centres for administration.
- Remember city sectors can be **multifunctional**.
- What examples do you have of how city sectors have changed over time?

For your UK city case study you need to be able to describe its structure:
- CBD
- inner city
- suburbs
- urban–rural fringe.

People travel to Birmingham's Bullring Shopping Centre from all over the UK. Located in Birmingham's CBD, the Bullring was redeveloped in 2003 at a cost of £50 million. 38 million visitors come to the Bullring each year.

How do the different parts of your city look?
- Where would you find the oldest buildings or the newest housing?
- How does population density change as you move through the city sectors?
- What is environmental quality like in different sectors; where would quality be lowest and highest?
- What evidence is there of land-use change in your city – for example, greenfield development on the urban–rural fringe, or brownfield development in the inner city?

Worked example

State **two** characteristics of the urban–rural fringe. **(2 marks)**

Land use is extensive because there is more space for expansion so, for example, retail centres have very large car parks.

There is a high residential population. This is an area dominated by expensive family housing.

This answer understands what is meant by 'characteristics'. Make sure you do more than say where the urban–rural fringe is located.

Now try this

Birmingham's history was as a centre of metalworking – since the Middle Ages. Name an industry or industries that your case study city was connected with in its past. **(1 mark)**

Had a look ☐ Nearly there ☐ Nailed it! ☐

Urban change differences

Case study

Changes happen in different ways in different parts of the city. This leads to inequalities between different parts of the city – for example, in education, health and employment. You need to know the reasons for these inequalities.

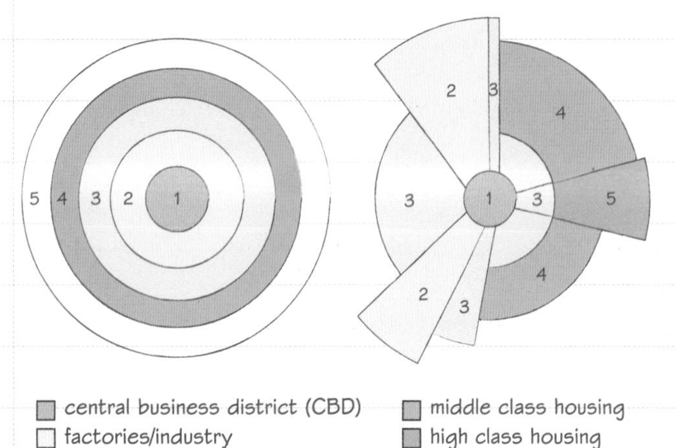

Different models have been developed to explain urban change. Here are three key ideas.

1. New arrivals to the city move to the cheapest areas in the inner city.
2. More established residents of the city move to the suburbs, where housing is more expensive but living conditions are better.
3. Industry and housing often develop along main roads and rail lines – this causes the wedges in the second diagram.

The Index of Multiple Deprivation

Deprivation means not having access to the same resources and opportunities as other people.

The **Index of Multiple Deprivation (IMD)** scores small areas across the whole UK for a range of different measures. All the areas are ranked from 1, the most deprived in the UK, to 32482: the least deprived area.

City change and migration

In 2011, 42 per cent of Birmingham's population was from an ethnic minority population, many from Pakistan and India. People come to Birmingham now from many places and for many reasons. New arrivals can feel more at home among communities of people from their old country.

- 40 per cent of Birmingham's residents live in areas described as among the most deprived 10 per cent in England.
- The areas of deprivation are found in a ring around the city centre.

Why do cities have deprivation 'hot spots'? Reasons include:

- a lack of jobs as industry moves out of the inner city, leaving behind those residents who cannot afford to move
- the inner city areas have old housing, which residents cannot afford to maintain: lowering environmental quality.
- crime increases with deprivation, reducing investment in the inner city.

Worked example

Identify two ways in which recent immigrants to a city impact on the character of city areas. **(2 marks)**

Immigrants can bring new cuisines with them and that can mean new types of restaurants and new types of food in shops and supermarkets.

New immigrants may strengthen the small business sector by increasing the number of shops and widening the types of services provided in the area.

Now try this

Identify an area of your case study city where deprivation is higher and an area of your case study city where deprivation is lower. Explain **two** differences between the two areas. **(4 marks)**

City challenges and opportunities

As the city changes it creates different challenges and opportunities – for example, regeneration of deindustrialised inner city areas.

Decline and decentralisation

In the second half of the 20th century, many UK cities went into decline: losing population. This was because of **deindustrialisation** – industries moving out of cities to cheaper locations.

Decentralisation also occurred. Land was cheaper and more space was available in the suburbs, so out-of-town shopping centres and business parks developed away from the urban core.

Worked example

Study the OS 1:25 000 map extract below, which includes the Birmingham Business Park.
Identify **two** opportunities that this city location offers for a business park development. **(2 marks)**

This location on the urban–rural fringe has excellent transport links (a motorway and major roads) and plenty of flat land for business park expansion.

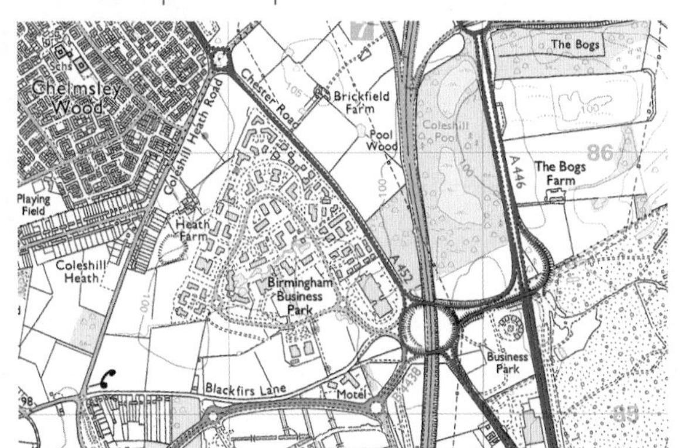

Birmingham's National Exhibition Centre (NEC) brings £2 billion to the regional economy each year.

Economic and population growth

Growth in cities occurs because economic opportunities attract people to live and work in city areas.

- **Sprawl on the rural–urban fringe** is driven by demand for housing, especially more expensive housing with a rural character.
- While manufacturing industry has declined in cities, **financial and business services** have expanded to replace lost jobs in the CBD and inner city. Globalisation means finance TNCs locate in major UK CBDs.
- **Gentrification** occurs when wealthier people move into deprived city areas where property is cheap. As these areas get more investment, poor residents are pushed out as rents and house prices rise.
- **Studentification** occurs in cities with universities when large numbers of young people become residents in student accommodation or other houses and flats. This can mean changes to land use with the construction of university accommodation blocks, and also social and cultural changes resulting from areas of the city becoming dominated by students.
- **Culture and leisure:** city authorities and private investors have built landmark cultural buildings and leisure facilities in city centres and on city outskirts.

The Selfridges Building in Birmingham's CBD is a city landmark that gives Birmingham a modern, exciting image. This encourages more investment in the city.

Now try this

Identify **one** example for your case study city of decentralisation or one example of growth/expansion. **(1 mark)**

Can you find these two examples on OS maps of your city?

Had a look ☐ Nearly there ☐ Nailed it! ☐

Improving city life

Case study

There are different strategies that can improve ways of life in a city. Regeneration schemes can redevelop deprived areas, while environmental initiatives can improve quality of life.

Positive impacts of regeneration

- New job opportunities in the area.
- Residents have better access to services – new retail outlets, cinemas, leisure services.
- Derelict buildings get repurposed or rebuilt.

Negative impacts of regeneration

- The area becomes too expensive for poorer residents to live in.
- New jobs may be low-paid service jobs – bar work, café work – not higher-wage skilled jobs.
- Regeneration strategies have been similar in many UK cities; cities lose individuality.

Affordable housing

City authorities run **affordable housing** schemes.

- Property developers must include a set number of affordable homes when they are building new housing estates.
- Landlords who rent property also work with the city authorities to make low-rent housing available to poorer people.
- City authorities provide support to help people find affordable housing.

Sustainable cities are more pleasant places to live in: greener, less polluted, easier to get around in and less expensive to live in.

Ways to make urban living in the UK more sustainable

Use brownfield sites for development to:
- improve appearance of these areas
- create new green spaces in the city.

Reduce waste by:
- recycling – 90% of household waste is recyclable
- reusing, e.g. bottles, plastic bags, etc.

Renovate old buildings to:
- enhance the appeal of the area
- improve energy efficiency.

New housing that is:
- affordable to rent or buy
- energy efficient.

Improve public transport systems by:
- linking bus, tram and rail routes
- providing feeder services to housing estates
- using environmentally friendly vehicles.

Involve communities in local decision making:
- consult local people instead of imposing plans
- put people first; ask for and act on their ideas
- foster the growth of a community spirit.

Worked example

Explain two ways in which quality of life in urban areas can be improved. **(4 marks)**

Investing in public transport, introducing congestion charging and pedestrianising shopping streets significantly improves urban quality of life.

Encouraging developers to include entertainment facilities like theatres and restaurants in new CBD retail developments improves quality of life because it means these areas are used at night as well as in the day.

Now try this

Describe **two** examples from your case study city of how quality of life has been improved. **(3 marks)**

The city and rural areas

The city and the rural areas around it (accessible rural areas) are linked together so that changes in the city affect the rural areas too.

There are flows between rural and urban areas.

migration
- rural–urban: for jobs, lifestyle, education
- urban–rural: for quality of life (commuters)

services
- rural residents use urban hospitals, universities
- urban residents use rural areas for recreation

goods
- rural residents get consumer goods from cities
- urban residents get food from rural areas

money flows between urban and rural areas
people flow between urban and rural areas

The relationship between urban areas and accessible rural areas has costs and benefits.

- 👍 Economic: people living in rural areas but working in the city can get higher wages.
- 👍 Social: people in accessible rural areas can access urban centralised services, like hospitals.
- 👍 Environmental: urban residents can access pleasant rural landscapes for recreation.
- 👎 Economic: urban decentralisation as developments move to the rural–urban fringe.
- 👎 Social: rural locations become too expensive for local young people to buy homes.
- 👎 Environmental: development of greenfield sites makes rural areas more urban.

Commuter villages

Features of an expanding commuter village
- richer newcomers and poorer locals
- many young families
- many older people who have retired from the city

You need to know why a rural area has experienced economic and social changes.

Pressure on housing. People moving from the city can afford to pay more for houses. This pushes up house prices so that there is less housing available that local people can afford.

Population change. Young people move from rural areas to the city; older people move from the city to rural areas.

Change in services. There has been a growth in recreation and leisure services in rural settlements but fewer people are going into farming jobs. Rural services like banks, post offices, pubs and shops have closed in many villages.

Commuter lifestyles. Because commuters leave early and get back late, commuter villages can seem empty during the week, only to become congested again at the weekends.

Worked example

Explain **two** ways in which cities and their surrounding rural areas are interdependent. **(4 marks)**

Cities that use renewable energy can be dependent on rural areas for power – for example, from wind farms or bioenergy.

Rural areas depend on cities for financial services like banks: they might have local cashpoints, but the banks that run them are in cities.

Now try this

For the rural area you have studied for your case study, describe **three** ways in which changes in the city have caused changes in the rural area. **(3 marks)**

Had a look ☐ Nearly there ☐ Nailed it! ☐

Rural challenges and opportunities

Case study Challenges in changing rural areas include problems like finding a job for young people or accessing healthcare for older people. Changing rural areas also bring opportunities to do something different from farming – rural diversification like farm shops, bed and breakfast, tourism and leisure activities.

Worked example

Study the map extract opposite, which is of a rural area 4 km south of Birmingham.

(a) Which of the following features on the map is most likely to be an example of rural diversification? **(1 mark)**

- ☐ **A** North Worcestershire Path
- ☐ **B** pits (disused)
- ☐ **C** cricket ground
- ☒ **D** Kings Norton Golf Club

(b) State **two** reasons why changing rural areas bring opportunities for rural diversification. **(2 marks)**

People moving from cities to rural areas increases the market for services like golf courses. People commuting to the city from rural areas can access higher wages. Then they spend more money on leisure activities where they live.

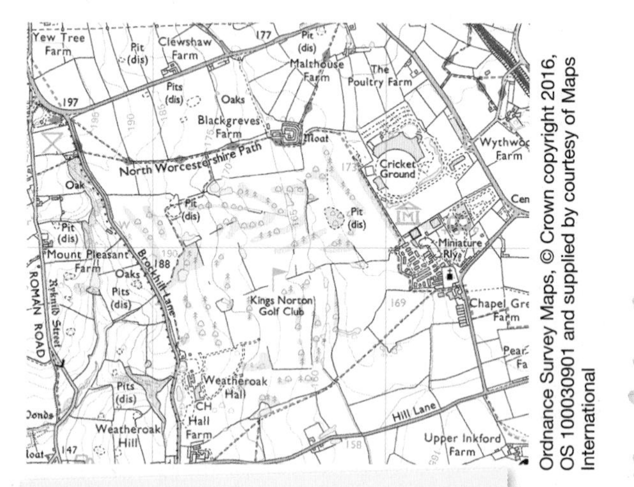

For a question like this you just need to state two valid reasons.

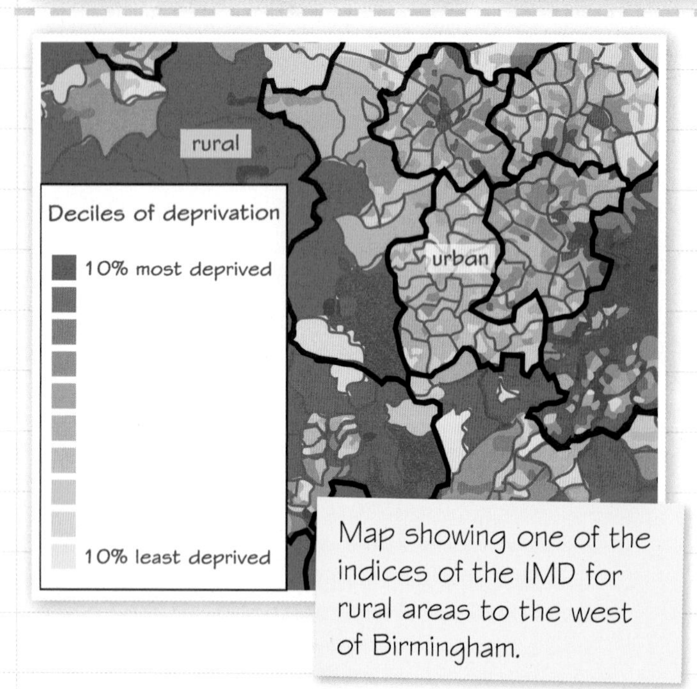

Map showing one of the indices of the IMD for rural areas to the west of Birmingham.

Deprivation differences

Most of the IMD indices show that city areas are more deprived than rural areas – for example, income deprivation (what people earn), the percentage of working people claiming benefits, the percentage of people suffering ill health and the percentage of young people underachieving in education.

The map opposite is for **barriers to housing and services.** These rural areas have higher scores for deprivation than the urban areas because of problems accessing affordable housing and 'geographical barriers': distances to services like doctors, petrol stations, post offices, shops and schools.

Now try this

Study the map extract at the top of this page. Suggest **two** ways in which rural diversification could have environmental impacts on this rural area. **(2 marks)**

Investigating dynamic urban areas: developing enquiry questions

Enquiry questions are the kind of questions that can be investigated by fieldwork in urban environments. They give fieldwork a purpose. You will have put together enquiry questions for your fieldwork.

In your exam you will be asked questions about the fieldwork you did, and also some questions where you will need to apply what you learned on your fieldwork to new situations.

The enquiry process

There are six stages in the enquiry process and you will be asked questions on at least two of them in the exam.

- An enquiry question often relates to a geographical theory: the sort of theory that can be tested through fieldwork.
- Key questions/hypotheses follow from the enquiry question, and they can be tested.

For example, an enquiry question could be:

- How does urban quality of life vary within inner city residential locations in [city name]?

A key question following on from this could be:

- Does environmental quality vary across five inner city residential sites in [city name]?

Geographical examples and theories

You need to be able to identify the key geographical concepts that the investigation is based on.

For example, for the enquiry question: **How does urban quality of life vary within inner city residential locations in [city name]?** geographers would use theories about urban change that help explain why inner city areas often experience issues of deprivation:

- issues with education and skills
- issues with employment
- issues with crime and anti-social behaviour.

Worked example

Study this map opposite, which shows levels of multiple deprivation in wards of Brighton in 2010. Students at the school marked on the map decided that ward 6 would not be an appropriate place to carry out fieldwork on variations in quality of life in Brighton's inner city wards. Explain one reason why this ward was not appropriate. **(2 marks)**

Ward 6 is a relatively large ward, which would have made carrying out a survey there very time-consuming.

There are several other reasons why ward 6 might not have been appropriate – for example, ward 3 has the same level of deprivation as ward 6 and is of a comparable size to other inner city wards. However, only one reason was required.

Now try this

Describe the location of your urban fieldwork. Explain why it was a good place to investigate how and why quality of life varies within urban areas. **(4 marks)**

Had a look ☐ Nearly there ☐ Nailed it! ☐

Investigating dynamic urban areas: techniques and methods

You will have used several different fieldwork techniques and methods in your investigation. You need to know what these techniques and methods are appropriate for, and what things to watch out for when using them to avoid making errors in the field.

Worked examples

Explain **one** reason why the method you used to measure environmental quality in the urban area you studied was appropriate to the task. **(2 marks)**

Name of method used: An environmental quality survey

It was appropriate because using the same survey questions between sites allowed me to make meaningful comparisons.

You have used census data in your investigation. Explain **one** way using census data supported your investigation. **(2 marks)**

The census data gave us information such as the percentage unemployed in each ward, and the percentage with no qualifications or with degree level qualifications. Using this information helped us select wards with contrasting socio-economic profiles to include in our investigation.

Explain **one** possible source of error when you collected data on perceptions of quality of life. **(2 marks)**

One possible source of error was my sampling strategy. I find it easier to talk to people who are about the same age as me so I didn't ask many older people to do the survey questionnaire. So the sample would be biased toward the perceptions of younger people and was therefore not 100% representative.

For the exam you need to know about:

- one **qualitative** fieldwork method to collect data on the views and perceptions of quality of life. Qualitative methods record descriptive data such as how people feel.
- one **quantitative** fieldwork method to collect data on environmental quality. Quantitative methods record data that can be measured as numbers.
- you need to be able to **say** why the method that you used was appropriate: there isn't just one correct method.

For the exam you need to know about two secondary data sources:

- census data such as the **Office for National Statistics** Neighbourhood Statistics
- one other source, which your teacher is likely to suggest for you.

Secondary data are data someone else has already collected. It will be useful for you to know about the different ways in which secondary data sources supported your investigation, and also about any particular advantages and disadvantages of your secondary data sources.

You may be asked questions that require you to reflect on the methods you used and consider any ways in which problems could have occurred. For the exam, it will be useful to revise possible disadvantages of your urban fieldwork methods.

Now try this

One indicator of urban quality of life is 'burglability' – how secure an area is. Name **three** things you would look for in an urban area to measure 'burglability'. **(3 marks)**

Investigating dynamic urban areas: working with data

You need to know about ways to process and present fieldwork data, how to analyse it and how to make conclusions and summaries backed up by evidence from the data.

Worked example

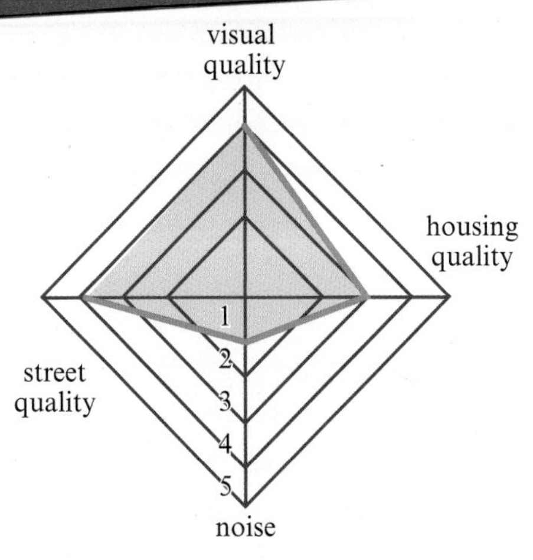

A student investigated variations in environmental quality in different urban locations. She presented her data using a radar graph, one graph for each location. Explain **one** advantage of using a radar graph to present these data. **(2 marks)**

An advantage of a radar graph is that it can display data on several different variables so it is a good way to compare the different characteristics of an area.

This answer has correctly identified one advantage of radar graphs.

The exam may ask questions about how the presentation of fieldwork data could be improved. Use your knowledge of what the advantages and disadvantages are of different types of data presentation.

Data presentation advantages

This is a list of the top five disadvantages. You might be able to think of others, too.

1. Isopleth maps: ideal for showing gradual change over an area.
2. Proportional symbols map: very accessible and easy to understand.
3. Kite diagrams: show changes over distance (e.g. transect data).
4. Dot maps: give a clear indication of differences in density for a geographic distribution (e.g. tourist signage).
5. Flow maps: show direction and volume of movement (e.g. for vehicle counts).

Using GIS (geographical information system)

Advantages of using GIS (if available):

- Great for showing spatial changes
- Different options and tools help with data analysis
- Can speed up data presentation

Now try this

The student who used the radar graphs to present her data on environmental quality used a separate radar graph for each of the five sites in her survey. Suggest **one** disadvantage of her presenting her data this way. **(1 mark)**

Familiar and unfamiliar

Some of the questions in the fieldwork sections of your Geographical investigations (Paper 2) exam will ask about the fieldwork that you did.

Questions for this 'familiar' fieldwork will ask things like: 'Explain one reason why the method you used…'.

Other questions will be about fieldwork that is unfamiliar to you. Here you need to apply what you know to fieldwork done by other people. These questions will often start by saying what 'A group of students' did in their fieldwork.

Investigating changing rural settlements: developing enquiry questions

Enquiry questions are the kind of questions that can be investigated by fieldwork in rural environments. They give fieldwork a purpose. You will have put together enquiry questions for your fieldwork.

In your exam you will be asked questions about the fieldwork you did, and also some questions where you will need to apply what you learned on your fieldwork to new situations.

Enquiry questions in your exam

There are six stages in the enquiry process and you will be asked questions on at least two of them in the exam.

- An enquiry question often relates to a geographical theory that can be tested through fieldwork.
- Key questions/hypotheses follow from the enquiry question, and they can be tested.

For example, an enquiry question could be:

- How does quality of life vary within rural settlements?

A key question following on from this could be:

- Does environmental quality vary between two parishes in [name of county]?

Geographical examples and theories

You need to be able to identify the key geographical concepts that the investigation is based on. For example, for the enquiry question: '**How does quality of life vary within rural settlements?**', the key geographical concepts could include:

- issues with a decline in rural services
- issues with rural housing prices due to second home purchasing
- issues with out-migration of rural young people
- issues with low wages compared to urban areas.

Worked example

Study the map opposite, which shows levels of multiple deprivation in Herefordshire parishes in 2015.

Students at a school in parish 1 decided that parish 4 would not be an appropriate place to carry out fieldwork on variations in deprivation in Herefordshire. Explain one reason why this parish was not appropriate. **(2 marks)**

Parish 4 is a relatively large parish, which would have made carrying out a survey there very time-consuming.

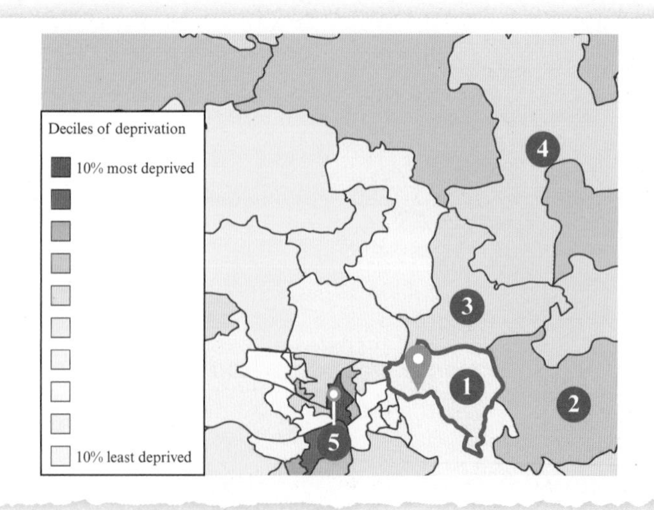

There are several other reasons why parish 4 might not have been appropriate – for example, it is a long way from the location of the school and it is at the same level of deprivation as parish 3.

Now try this

Describe the location of your rural fieldwork. Explain why it was a good place to investigate how and why quality of life varies within rural areas. **(4 marks)**

Investigating changing rural settlements: techniques and methods

You will have used several different fieldwork techniques and methods in your investigation. You need to know what these techniques and methods are appropriate for, and what things to watch out for when using them to avoid making errors in the field.

Worked examples

Explain **one** reason why the method you used to measure environmental quality in the rural area you studied was appropriate to the task. **(2 marks)**

Name of method used: An environmental quality survey

It was appropriate because using the same survey questions between sites allowed me to make meaningful comparisons.

You have used census data in your investigation. Explain **one** way using census data supported your investigation. **(2 marks)**

The census data gave us information such as the percentage unemployed in each parish, the percentage with no qualifications or with degree level qualifications. Using this information helped us select parishes with contrasting socio-economic profiles to include in our investigation.

Explain **one** possible source of error when you collected data on perceptions of quality of life. **(2 marks)**

One possible source of error was my sampling strategy. I find it easier to talk to people who are about the same age as me so I didn't ask many older people to do the survey questionnaire. So the sample would be biased toward the perceptions of younger people and not 100% representative.

Qualitative and quantitative

For the exam you need to know about:

- ✓ one qualitative fieldwork method to collect data on the views and perceptions on quality of rural life. Qualitative methods record descriptive data, such as how people feel
- ✓ one quantitative fieldwork method to collect data on environmental quality. Quantitative methods record data that can be measured as numbers.

You need to be able to say why the method that you used was appropriate; there isn't just one correct method.

For the exam you need to know about two secondary data sources:

- census data such as the **Office for National Statistics** Neighbourhood Statistics
- one other source, which your teacher is likely to suggest for you.

Secondary data are data someone else has already collected. It will be useful for you to know about the different ways in which secondary data sources supported your investigation, and also about any particular advantages and disadvantages of your secondary data sources.

You may be asked questions that require you to reflect on the methods you used and consider any ways in which problems could have occurred. For the exam, it will be useful to revise possible disadvantages of your rural fieldwork methods.

Now try this

One indicator of rural quality of life is 'burglability' – how secure an area is. Name **three** things you would look for in a rural area to measure 'burglability'. **(3 marks)**

Had a look ☐ Nearly there ☐ Nailed it! ☐

Investigating changing rural settlements: working with data

You need to know about ways to process and present fieldwork data, how to analyse it and how to make conclusions and summaries backed up by evidence from the data.

Worked example

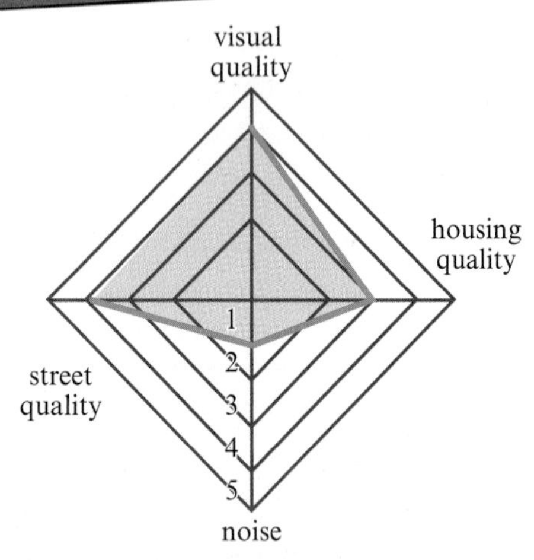

A student investigated variations in environmental quality in different rural locations. She presented her data using a radar graph, one graph for each location.
Explain **one** advantage of using a radar graph to present these data. **(2 marks)**

An advantage of a radar graph is that it can display data on several different variables so it is a good way to compare the different characteristics of an area.

The answer explains a valid advantage.

Advantages and disadvantages

The exam may ask questions about how the presentation of fieldwork data could be improved. Use your knowledge of what the advantages and disadvantages are of different types of data presentation.

Data presentation advantages

A top five list – there are others...

1. Isopleth maps: ideal for showing gradual change over an area.
2. Proportional symbols map: very accessible and easy to understand.
3. Kite diagrams: show changes over distance (e.g. transect data).
4. Dot maps: give a clear indication of differences in density for a geographic distribution (e.g. for tourist signage).
5. Flow maps: show direction and volume of movement (e.g. for vehicle counts).

Using GIS (geographical information system)

Advantages of using GIS (if available):

- great for showing spatial changes
- different options and tools help with data analysis
- can speed up data presentation.

Now try this

The student who used the radar graphs to present her data on environmental quality used a separate radar graph for each of the five sites in her survey. Suggest **one** disadvantage of her presenting her data this way. **(1 mark)**

Familiar and unfamiliar

Some of the questions in the fieldwork sections of your Geographical investigations (Paper 2) exam will ask about the fieldwork that you did.

Questions for this 'familiar' fieldwork will ask things like: 'Explain one reason why the method you used...'.

Other questions will be about fieldwork that is unfamiliar to you. Here you need to apply what you know to fieldwork done by other people. These questions will often start by saying what 'A group of students' did in their fieldwork.

Paper 2

Paper 2 is UK Geographical Issues. It has three sections: A, B and C. Each contains a range of different types of question, including an 8-mark extended writing question.

The three sections of Paper 2 are:

- A: the UK's evolving physical landscape
- B: the UK's evolving human landscape
- C: geographical investigations

In the extended writing question in Section C you will apply what you learned from your own fieldwork to a new situation.

SPGST

In Section A of Paper 2, the extended writing question will be worth 8 marks for the extended writing answer, plus 4 marks for SPGST = 12 marks in total.

It will always be very clear which of the extended writing questions has marks available for SPGST.

Assessment Objective 4

In Paper 2 the 8-mark questions involve the command words 'Assess' or 'Evaluate'.

- 4 marks are for Assessment Objective 3
- 4 marks are for Assessment Objective 4

AO3 is **applying** your knowledge and understanding to geographical issues, using evidence to come to a judgement.

AO4 is **selecting** the right skill or technique to investigate the question and to communicate your answer.

What to aim for

A top answer will show that you can:

- apply your understanding to unpick the different factors involved in the question
- put together a clear, logical argument
- use evidence to decide which of these factors are more important than others
- use your geographical skills to get accurate information.

Extended writing questions

Study this 8-mark Section A question:

Analyse the diagram opposite, which indicates the ways in which UK urban areas have experienced changes over the last decade.
Assess the causes of variations in urban growth or decline in the UK. **(8 marks)**

Use your geographical understanding to analyse the different reasons why cities grow or decline (AO3). Pull out information from the diagram that supports your analysis (AO4).

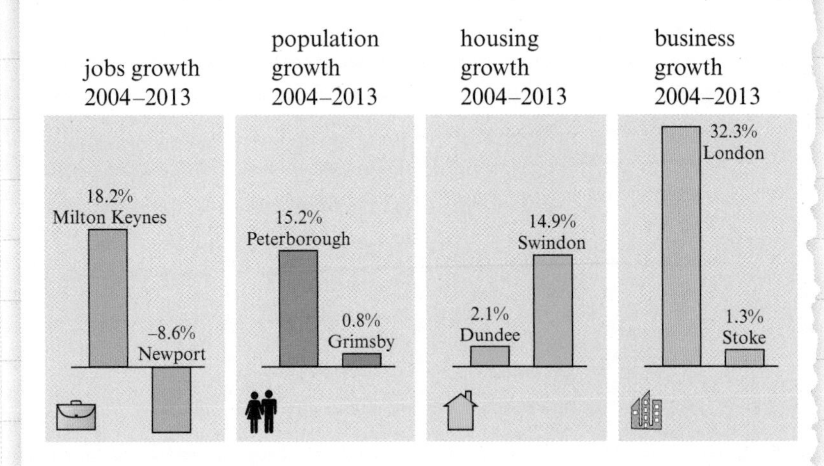

Now try this

Try the 8-mark Section A question opposite for yourself. Remember that 'assess' means that you need to use evidence to weigh up the different factors involved, and come to a judgement as to which is the most important.

Distribution of major biomes

You need to be able to define the terms **ecosystem** and **biome**. You also need to know the location of the world's biomes and how they are connected to climate.

Definitions

Ecosystem – a grouping of plants and animals that interact with each other and their local environment

Biome – a large ecosystem; a grouping of plants and animals over a large area of the Earth

Taiga biome

Taiga (boreal) forests are at higher latitudes where the Sun's rays are weak. Trees are adapted to the cold with needle-like leaves.

Temperate biome

Temperate forests have high rainfall and there are seasonal variations in the Sun's rays. Trees lose their leaves in the cool winters.

Tundra biome

The **tundra** is within the Arctic Circle. The Sun gives little heat here and there is little rainfall. Only tough, short grasses survive.

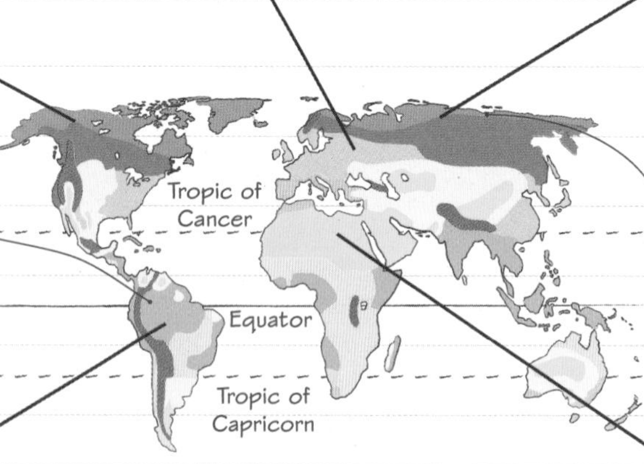

Iquitos: °C 0–40; mm 0–500; J F M A M J J A S O N D; Month

Kazachye: °C −40–20; mm 0–120; J F M A M J J A S O N D; Month

tropical forest
tropical grasslands
desert
mediterranean
temperate forest
mountain
temperate grassland
taiga forest
tundra

Tropical biome

Tropical rainforests are mostly found either side of the Equator. The temperature is hot and there is heavy rainfall.

Desert biome

Deserts are close to the tropics of Cancer and Capricorn. This is where hot dry air sinks down to the Earth's surface and the Sun's rays are concentrated making it very hot in the day.

Worked example

Explain one way in which climate influences the distribution of a major biome. **(2 marks)**

The high temperatures and precipitation of the tropical rainforest biome produce very high biodiversity. Tropical rainforest is found within 10 degrees north and south of the equator. The Sun's rays hit the equator straight on, which gives the tropical rainforest zone high average annual temperatures (above 20°C all year).

Most students know a lot about the tropical rainforest and the taiga. It is always a good idea to write about something you know about – just be careful to stick to what the question is asking for.

Now try this

1 Which of the biomes shown on the map on this page has the **lowest** number of hours of sunshine per year? **(1 mark)**

☐ **A** desert
☐ **B** mediterranean
☐ **C** taiga (boreal) forest
☐ **D** tundra

2 State two differences between the climate graphs for the tropical rainforest and tundra biomes shown on the map. **(2 marks)**

Local factors

The distribution of major biomes is explained by climate, but there are also some local factors that affect where biomes are found: **altitude**, **rock and soil type**, and **drainage**.

Worked example

Study the picture shown here. Explain why fewer plants grow along the river. **(1 mark)**

Poor drainage along the river means waterlogged conditions that are not suitable for most plants.

This is a good answer because it identifies a correct reason (poor drainage) and then explains why it affects vegetation growth.

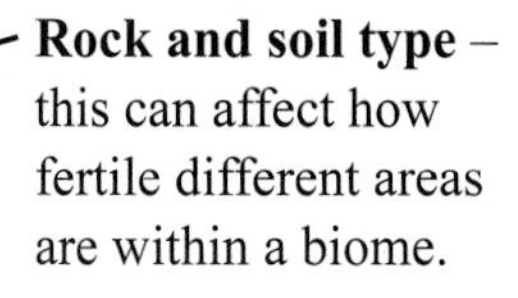

Altitude – different plants grow at different temperatures within the same biome. The higher the altitude the lower the temperature.

Rock and soil type – this can affect how fertile different areas are within a biome.

Drainage – swamps and bogs occur where drainage is poor. Fewer, more specialist plants grow in boggy areas.

Biotic and abiotic components

Biotic and abiotic components of an ecosystem **interact**. For example:

The taiga biome has low **biodiversity**.

↓

Abiotic components – long cold winters, low precipitation, frozen soils

↓

Biotic components – only specialist plants that can tolerate poor soils, low light, cold temperatures; small amounts of plant food = small numbers of animals

Definitions

Biotic – the **living** components of an ecosystem: the plants (flora) and the animals (fauna)

Abiotic – the **non-living** components of an ecosystem – for example, soils, rocks, water, the atmosphere

Biodiversity – the variety of biotic components in an ecosystem (high biodiversity = thousands of different plants and animals)

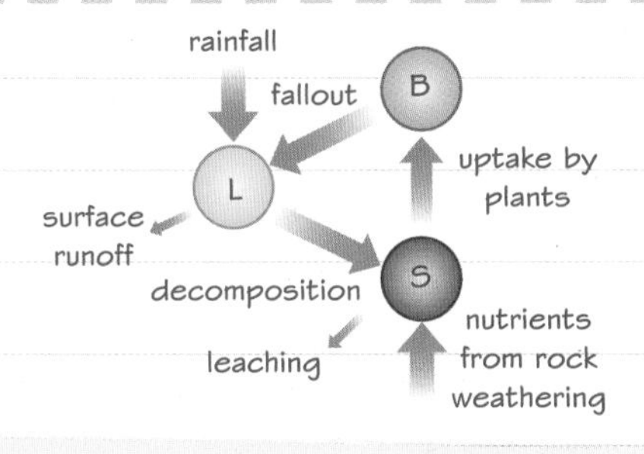

Nutrient cycles show how biotic and abiotic components interact. L = Litter store, B = Biomass store and S = Soil store. The size of the arrows shows the size of the nutrient transfer between each store.

Now try this

This photo is from the tropical rainforest biome in Brazil. Identify one biotic factor influencing the local distribution of tropical rainforest shown in this photo. **(1 mark)**

Had a look ☐ Nearly there ☐ Nailed it! ☐

Biosphere resources

The **biosphere** provides humans with some of our most essential resources: the food we eat and many of our medicines, building materials and sources of fuel.

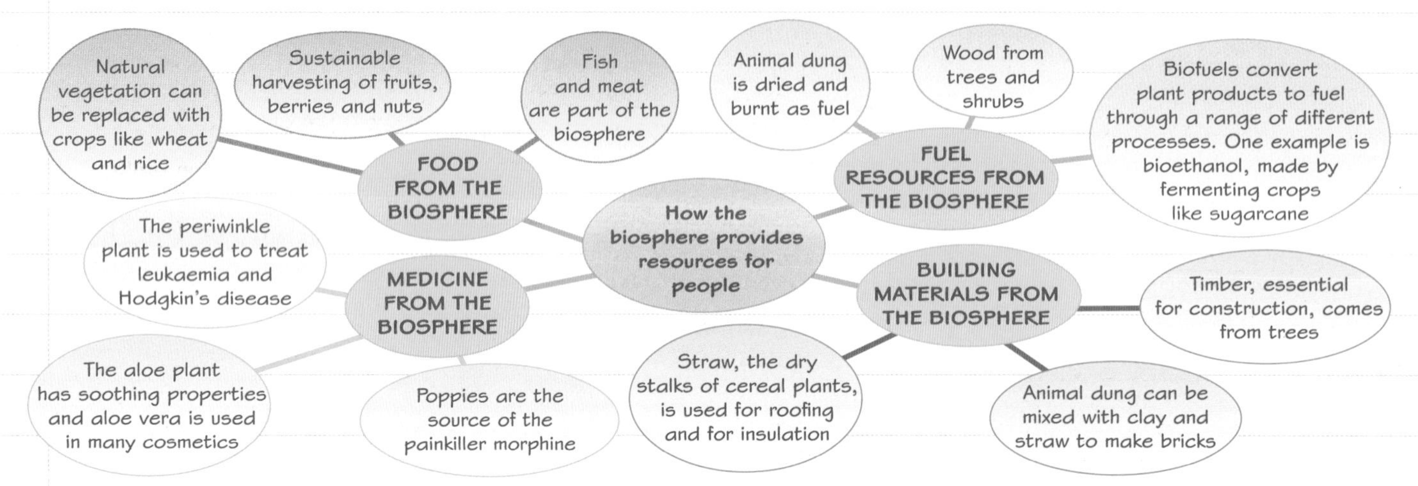

The biosphere provides resources for indigenous and local people: food, medicine, building materials and fuel resources. In developed countries few people now use resources directly from the biosphere.

The biosphere is increasingly exploited for its resources – for example, the demand for some fish species has led to overfishing and huge declines in fish numbers.

Worked example

Study this satellite image of the Athabasca Oil Sands mine in the taiga biome of Canada. Explain the impacts this mining operation could have on biosphere services for local people. **(4 marks)**

The mine is extensive (the scale suggests around 20 km²) and it is located next to the Athabasca River. A large area has been disturbed, destroying plants and animals that may have provided food, fuel, building materials and medicines for local people. Because the mine is extracting oil, there is a high risk of water and air pollution. Drinking water is a vital biosphere resource; if the river became polluted that could have a very negative impact on local people.

Always check whether satellite images have a scale. If they do, you can use the scale to make your answer more precise.

Now try this

A decline in the number of orangutans in Malaysia and Indonesia has been linked to the increase in plantations for the biofuel palm oil. This is an example of commercial exploitation of the biosphere for which of the following? **(1 mark)**

☐ energy resources ☐ water resources ☐ mineral resources

Biosphere services

The biosphere provides essential services for all life on Earth. It is a life-support system for the planet.

A life-support system

Biosphere – what it does for us

- It regulates the gases that make up the atmosphere – plants absorb carbon dioxide and produce oxygen for us to breathe in.
- It regulates the water cycle – plants slow the flow of water to rivers and filter water to make it clean.
- It keeps soil healthy for plants to grow – new nutrients are provided by rotting plant material.

Worked example

Study the leaflet opposite, which is from a charity campaigning for the protection of the rainforest in South America.

Using evidence from the resource, explain how the biosphere regulates water within the hydrological cycle. **(8 marks)**

The hydrological cycle (water cycle) describes how water moves between different stores – for example, evaporation and then condensation form clouds, and the water returns to the surface as precipitation.

The biosphere is all living things on Earth. Vegetation is the living thing that has the most impact on the hydrological cycle.

Vegetation intercepts precipitation. So in the rainforest, vegetation reduces surface run-off. This is an important process because the heavy rainfall of equatorial climates would otherwise increase the risk of frequent flash flooding and landslides.

When rainforest is on hills it intercepts moisture directly from clouds. Instead of the clouds releasing their rain all in one go, the trees release the water gradually throughout the year. This helps guarantee water supplies for whole regions.

To summarise, a rainforest acts like a giant sponge, regulating the heavy precipitation of the equatorial climate.

5 REASONS TO PROTECT THE RAINFOREST

1. **Protecting biodiversity** The rainforest is the most biodiverse ecosystem on Earth. 7000 drugs have rainforest origins. How many more drugs are still to be discovered?
2. **Climate change** Plants soak up CO_2 from the atmosphere. The Amazon rainforest soaks up 2 billion tons of carbon a year.
3. The forest canopy **protects the soil** from being eroded by heavy rainfall. Without the trees, flash floods and landslides are more likely to occur.
4. **Cloud forests** strip water from the atmosphere and release it slowly throughout the year into the region's rivers and streams.
5. Some rainforests are home to peoples whose **traditional way of life** depends totally on the forest. Destroying the rainforest robs these people of their cultural identity.

The exam questions for this component will often ask you to apply your geographical understanding to new pieces of information, like this leaflet.

Now try this

Identify **two** reasons why humans should look after the biosphere. **(2 marks)**

Pressure on resources

The world's population has grown **exponentially**. More people means more demand for resources – food, energy and water. Other trends also increase resource demand: urbanisation, industrialisation and affluence.

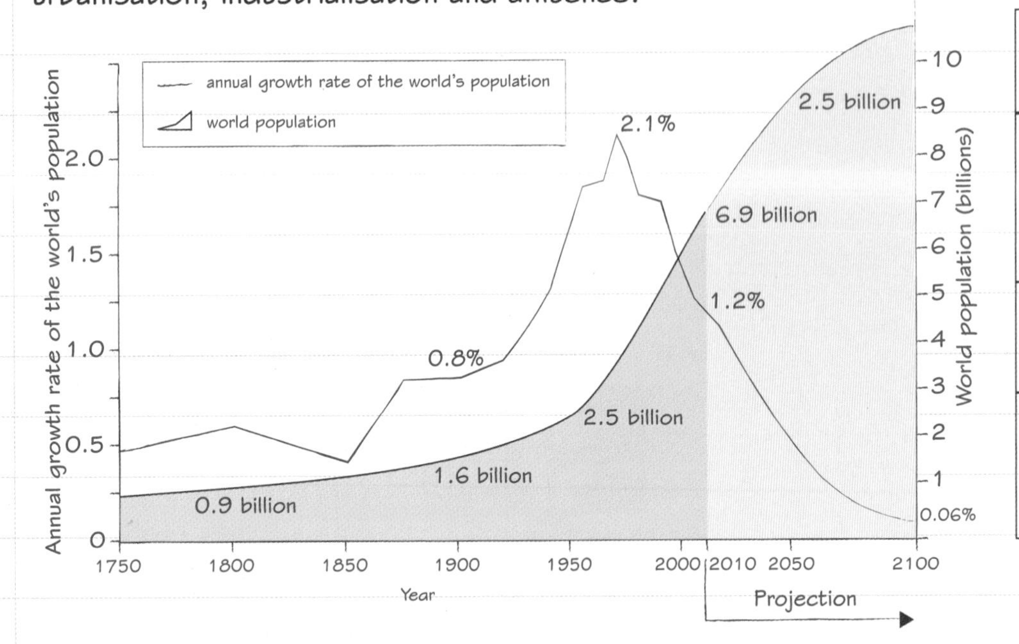

Urbanisation
The world is 54% urban – by 2050 this will be 66%

Food production
Current levels of food production will need to double by 2050 to feed the world's population

Water shortages
The demand for water will increase by 55% by 2050

Affluence
The number of middle class people will grow from 1.8 billion to 4.9 billion by 2030

Thomas Malthus (1766–1834)

Malthus' theory was that it was impossible to increase food production as rapidly as population growth. Therefore if a population was allowed to grow too much, it was inevitable that food supply would run out and famine would result, reducing the population size.

Ester Boserup (1910–99)

Boserup's theory was that human innovation will be sparked by demands on resources. So if there is a high demand for food resources, new techniques to increase food production will be invented. The same applies to water supplies (e.g. desalination) and energy supplies.

Worked example

In 2010, Chinaautoweb reported in an article that in 1990 there were 5.6 million vehicles on China's roads. By 2014 this was 264 million vehicles (104 million of which were cars).

By what approximate percentage has the number of vehicles in China increased between 1990 and 2014?

(1 mark)

☐ 46 per cent ☐ 460 per cent ☒ 4614 per cent

The vehicle statistics are a good example of the impact of growing affluence in China: more people have enough money to afford a vehicle. More vehicles means more demand on raw materials and energy.

To work out percentage increase, first take the smaller number away from the bigger number to find the difference between them. Then divide your answer by the smaller number and multiply by 100 to get your percentage.

Now try this

Explain how urbanisation or industrialisation increases demand for water resources. **(2 marks)**

Tropical rainforest biome

The tropical rainforest ecosystem has evolved and adapted to the equatorial climate, and its plants and animals are adapted to the challenges and opportunities of that climate.

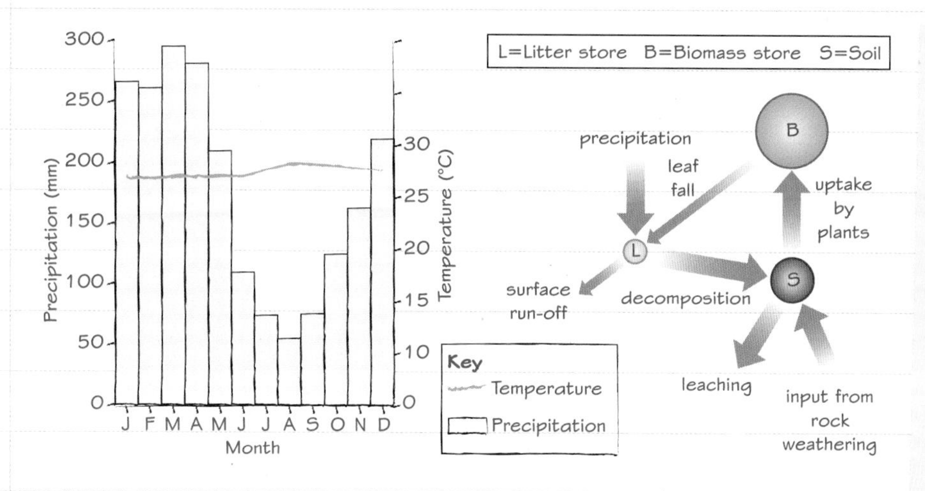

Biotic and abiotic components interrelate in all ecosystems. **Abiotic** includes climate, soil and water. **Biotic** includes plants, animals and humans. Climate graphs show climate characteristics; nutrient cycle diagrams illustrate interdependencies between biotic and abiotic characteristics.

Top three plant adaptations

1. **The dense forest canopy blocks out light.** Some trees, called **emergents**, grow 40 m tall, 10 m above the canopy.
2. **Mould grows on all wet surfaces.** This would block sunlight from leaves. Most plants have evolved **drip tips** that channel water off the leaf.
3. **Nutrients are concentrated in only the top layer of the soil.** This means tree roots have to be shallow. **Buttress roots** give the tall trees extra stability.

Rainforest layers

The tropical rainforest has five main layers: herb layer, shrub layer (including young trees), under-canopy, main canopy and emergent layer. Different plants and animals are adapted to each layer.

Worked example

Study this photo of a chameleon from a rainforest in Madagascar. Identify **two** ways in which this chameleon is adapted to the rainforest environment. **(2 marks)**

Many animals use camouflage to avoid predators in the rainforest. Chameleons can change their skin colour to match their surroundings. The chameleon has feet that are adapted to grip branches and a tail that it can wind around branches in the canopy.

Two valid ways are identified, with information on why they are adaptations to the rainforest.

Now try this

Pigs dig holes in the rainforest leaf litter and topsoil, which quickly fill with rainwater. Insects collect in these rainwater puddles, and the pigs return to the puddles to eat the insects.
Identify the abiotic and biotic components in this description. **(2 marks)**

Had a look ☐ Nearly there ☐ Nailed it! ☐

Taiga forest biome

The climate of the taiga biome is highly seasonal and extreme. The winters are long and very cold; the summers are warm and wet but very short.

Worked example

Study the climate graph opposite. Identify **three** ways in which the climate of the taiga is different from the equatorial climate of the rainforest biome. **(3 marks)**

In the tropical rainforest the temperature does not change much all year (variation of around 2°C). In the taiga the temperature is very variable: freezing cold winters and mild summers.

The taiga is much drier than the tropical rainforest. While some rainforests receive 2000 mm of rain a year, some taiga locations have only 450 mm a year.

The taiga has long, extremely cold winters when plants cannot grow. The tropical rainforest is above 20°C all year, so plants grow constantly.

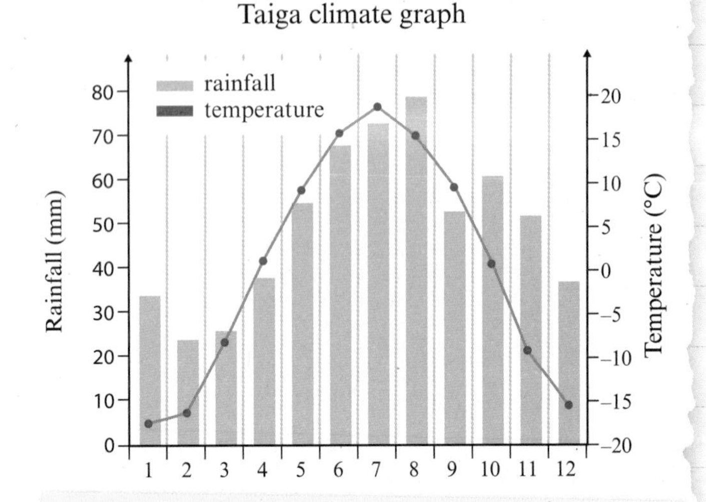

Remember that biotic and abiotic components interrelate in all ecosystems. Abiotic includes climate, soil and water. Biotic includes plants, animals and humans.

Taiga plants

Taiga plants have to deal with very low temperatures, a very short growing season and low-nutrient soils. Their adaptations include:

- **Needle-shaped** leaves: taiga trees do not drop their leaves. This is to maximise photosynthesis throughout the year. To reduce water loss the leaves are needle-shaped and waxy.
- **Cone-shaped**: many taiga trees have downwards facing branches to shed heavy snow.
- Few species – a **simple ecosystem structure**: few plants can deal with taiga extremes. **Coniferous** trees dominate, plus lichens and mosses. Trees grow close together to reduce wind damage.

Taiga animals

In the short taiga summer, huge numbers of insects attract many bird species. Most of the birds then **migrate** south for warmer winters.

The taiga supports some large mammals: both herbivores and carnivores.

- The moose (called an elk in Europe) is an example of a large taiga herbivore. It is one of the few animals that can eat pine needles.
- The brown bear is an example of a large taiga carnivore/omnivore. Bears build up fat layers in summer for hibernation in winter dens.

Non-migrating animal species often have coats or feathers that turn white in winter for camouflage and extra warmth.

Now try this

Study the photo of the moose on this page, which is licking salt off a road. Explain two negative impacts road building could have on the taiga biome. **(4 marks)**

Productivity and biodiversity

The tropical rainforest has very high biodiversity and very high productivity. The taiga has much lower biodiversity and much lower productivity.

Rainforest nutrient cycle

- Plants grow all year in huge numbers.
- Dead matter drops to the forest floor and decomposes quickly in the warm, wet conditions.
- Fast-growing plants take up the nutrients very quickly.
- The constant precipitation leaches nutrients down through the deep rainforest soil.

rainforest
B
precipitation
leaf fall
uptake by plants
L
S
surface run-off
decomposition
leaching
input from rock weathering

Taiga nutrient cycle

- Plants can only grow in the short summer: 3–5 months.
- Litter accumulates because decomposition only happens in summer.
- Soils are thin, low in nutrients and acidic.
- Plants grow very slowly due to short growing season and low-nutrient soil.

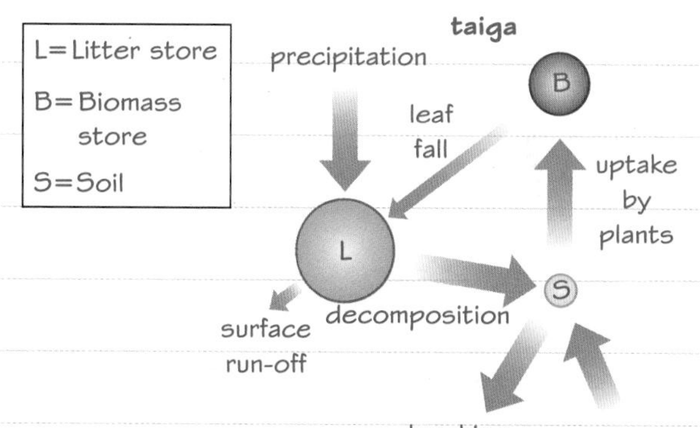

Worked example

Study the food web shown opposite. Explain why a taiga food web is much less complex than a food web for a tropical rainforest. **(4 marks)**

Biodiversity is much higher in tropical rainforests than in the taiga because the equatorial climate supports year-round plant growth. In the taiga, nutrient supply is much lower and can support far fewer plant and animal species: low biodiversity. In the taiga, the extreme winters mean that plant and animal species have to be specialised to survive. This means fewer species overall. In the rainforest there are many ecological niches, encouraging very high biodiversity.

A taiga food web from North America

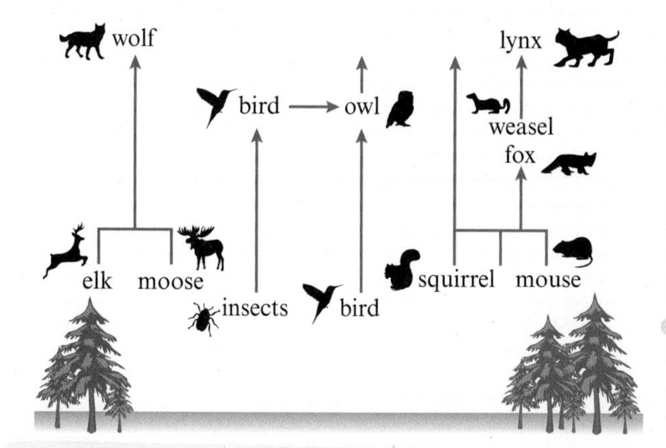

Make sure you understand the difference between biodiversity (the number of species) and productivity (a measure of the biomass the ecosystem can support).

Now try this

Using the nutrient cycle diagrams on this page to help you, explain **one** reason why most tree species in both the tropical rainforest and the taiga forest have shallow roots. **(2 marks)**

Had a look ☐ Nearly there ☐ Nailed it! ☐

Tropical rainforest deforestation

You need to know about some key causes of tropical rainforest deforestation and why climate change could threaten rainforest health.

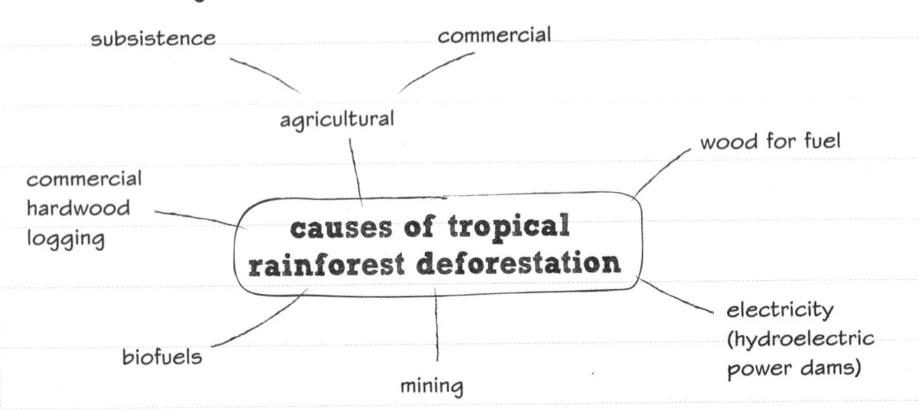

7.3 million hectares of rainforest are cleared each year: the equivalent of 36 football fields of forest being cleared every minute.

Exam questions could be on any of these causes of tropical rainforest deforestation, so be sure you know about all of them.

Worked example

The diagram above identifies different causes of rainforest deforestation. Choose **two** and explain how they lead to deforestation. **(4 marks)**

Biofuel can be produced from crops such as oil palm. Because of the high demand for biofuels internationally and because oil palms grow well in tropical climates, farmers can make a lot of money by clearing forest and planting oil palm plantations.

Subsistence agriculture is when poor people clear the forest to plant crops to feed themselves and their families. Rainforest soils quickly lose their nutrients when forest cover is cleared and also lose soil to surface run-off. This means subsistence plots are abandoned after a couple of years and new plots must be cleared from the forest.

Climate change threats

Global warming threatens the tropical rainforest.

- ✓ Warming global temperatures could cause a northward shift in the atmospheric system that brings constantly wet weather to tropical rainforests.
- ✓ The impact of this would make most tropical rainforests drier and hotter.
- ✓ Tropical rainforest plants and animals have evolved to constant temperature conditions: they cannot tolerate heat spikes.
- ✓ Tropical rainforest plants are not able to tolerate a long drought: it kills some and stresses the survivors.
- ✓ Stressed plants and animals have less resistance to disease.
- ✓ Drier forests are at risk of forest fires: the tropical rainforest ecosystem is not adapted to fire.

Now try this

The drainage of peat-swamp forest was one of the factors contributing to uncontrollable fires in Kalimantan in 1997–98. The fires lasted for 6 months and killed up to 8000 orangutans. Explain why climate change could also indirectly lead to more forest fires within the tropical rainforest biome. **(4 marks)**

Threats to the taiga

The taiga is under threat because of human activity – because of the ways humans use the taiga and also as a result of impacts from human activity in other biomes.

Direct threats to the taiga

The conifer trees of the taiga produce **softwood timber**. Most of the world's softwood comes from the taiga – half of it from Russia's taiga.

Direct threats to the taiga come from logging for softwood, pulp and paper production.

Indirect threats to the taiga

The taiga is also threatened by mining for minerals and fossil fuels (oil, gas), and when dams are created for hydroelectric power schemes, flooding forested valleys.

These are called **indirect threats** because damage to the taiga happens as a side effect.

Worked example

With reference to the diagram opposite, explain how acid precipitation threatens the taiga biome. **(4 marks)**

When power plants and factories burn coal or oil the smoke contains pollutants such as sulphur dioxide, CO_2 and nitrogen oxides that react with water vapour in the clouds and fall to the surface as acid precipitation. When prevailing winds carry pollution from industrial areas to taiga areas, the acid precipitation damages taiga trees and makes them less resistant to pests and diseases and less able to recover from forest fires.

Acid precipitation

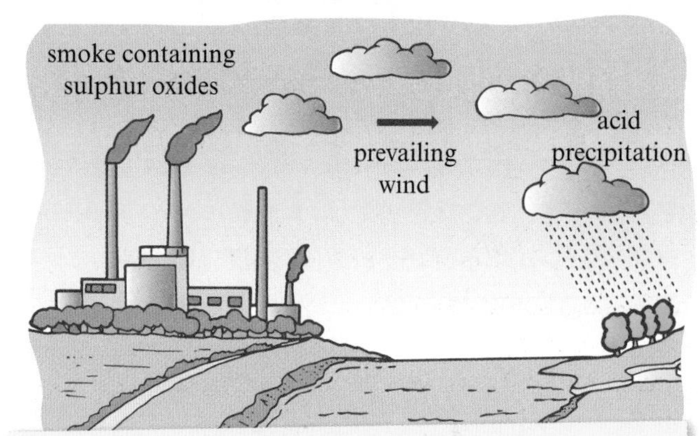

Do use Resource Booklet sources to answer questions – but avoid just repeating the information they provide.

Biodiversity at risk

Global warming as a result of industrial development also threatens the taiga.

- Animals like the Siberian tiger have heavy fur coats and high levels of body fat, making them heat intolerant.
- Warmer winter temperatures will allow new diseases and pests to spread to the taiga. Taiga animals and plants will not have resistance to these, so species could die out.
- Forest fires in Russia's taiga are 30–50 per cent more common than they were 20 years ago, which correlates with global warming. Taiga species are not adapted to frequent fires; new trees need many years to grow.

Now try this

In the UK, 78 per cent of the paper we used gets recycled. Recycling paper generates 70 per cent less CO_2 than making new paper. Explain **two** ways in which recycling paper helps reduce the impact of human activity on the taiga biome. **(4 marks)**

Had a look ☐ Nearly there ☐ Nailed it! ☐

Protecting the rainforest

Global actions to protect the tropical rainforest have advantages and disadvantages.

CITES

- **CITES** stands for the Convention on International Trade in Endangered Species of Wild Fauna and Flora.
- CITES currently protects 35 000 different species. Countries that sign up to CITES agree to stop exports or imports of endangered species.

REDD

- **REDD** stands for Reducing Emissions from Deforestation and Forest Degradation.
- REDD supports schemes that reduce the rate of deforestation.
- The United Nations monitors the schemes by the use of remote sensing and visits.

Advantages of CITES

- CITES has a huge international influence: 181 countries have signed up to it.

Disadvantages of CITES

- It is very difficult to check that all the countries are enforcing the CITES rules. For example, in 2014, over 1000 rhinos were killed by poachers in South Africa.

Advantages of REDD

- Because REDD is backed by the United Nations, very large sums of money are available for REDD projects. A REDD scheme in Brazil is backed by a US$1 billion fund.

Disadvantages of REDD

- It is not clear what REDD means by 'forest'. Some palm tree plantations received REDD funding, even though these damage rainforest.

Worked example

The graph opposite shows rates of deforestation in the Amazon rainforest of Brazil between 1988 and 2014.

Explain why deforestation rates are rising in some areas of the world but falling in others. **(4 marks)**

Brazil's deforestation rate has fallen since 2004. This is linked to government policies to reduce deforestation, such as withdrawing grants to clear rainforest land. Economic development in Brazil has made its exports more expensive and made farming less profitable.

Elsewhere, deforestation rates are high, especially in Africa. Here rapid population growth is combined with poor people using wood for fuel and clearing forest for farmland.

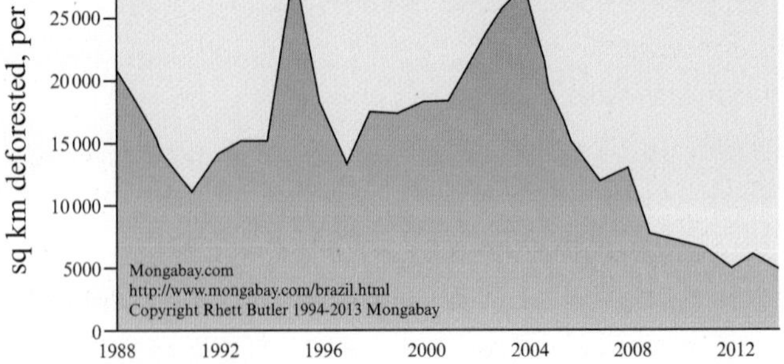

Deforestation in the Brazilian Amazon, 1988–2014

Now try this

Name **one** example of a rainforest 'service' that is threatened by deforestation. **(2 marks)**

Sustainable tropical rainforest management

Successful sustainable management has to make sure that forest conservation gives local populations real economic benefits.

What is sustainability?

Sustainability is the ability to keep something going at the same rate or level. There are several key ideas when considering this as a geographer. These are, that it:

- keeps going without using up natural resources
- doesn't require lots of money to keep it going
- meets the needs of people now and in the future without having a negative effect.

Sustainable biosphere management

- Ensures the ecosystem can recover quickly from any use.
- Prevents damage to the environment/ecosystem.
- Helps local people to benefit from their environment/ecosystem.
- Helps local people to understand why this management benefits them.

Possible tensions

1. **Economic** – individuals and communities often want to make as much money as possible, and may use the resources in the biosphere to do this. This provides tensions as it may damage or even destroy the environment in the long term.
2. **Social** – to be socially sustainable something must not benefit one group/individual at the expense of another, including future generations. It also means consulting people on an equal basis. If everyone is to benefit, this may put the environment at risk. There are also economic tensions as some businesses may flourish at the expense of others.
3. **Environmental** – being environmentally sustainable means not harming natural resources so they cannot regenerate or continue in the long term. This can conflict with making money and improving living standards for all.

Worked example

Most sustainable rainforest management schemes promote alternative lifestyles for local communities, such as ecotourism, forestry and sustainable agriculture. Describe **two** ways in which alternative lifestyles like these can help to protect remaining tropical rainforest areas from deforestation. **(4 marks)**

Ecotourism gives jobs to local people: as forest guides, in providing other services, and in selling local craft products to tourists. Because local people get money from tourists coming to see the rainforest, they want to protect the forest.

Sustainable agriculture helps protect the rainforest because it reduces the need for farmers to keep clearing new plots of land in the forest. Using better sources of fertiliser and better farming techniques enables farmers to keep using the same plots year after year and also to get better yields from their land.

Now try this

Explain **one** way in which sustainable forest management can be difficult to achieve. **(2 marks)**

Protecting the taiga

It is difficult to both protect the taiga and allow its resources to be used.

Why protect the taiga?

The taiga is a fragile ecosystem and takes a long time to recover from damage.

- Plants grow very slowly because of the lack of nutrients and because of cold winters. Pollution remains in the ecosystem for decades.
- There are very few species in the taiga. A disease that affects one species impacts the whole ecosystem.
- Taiga animals and plants are highly specialised. They will struggle to adapt to climate change.

National parks and protected wilderness

These areas prevent any exploitation of natural resources.

- **Conservation** protects plants and animals by looking after and restoring their natural habitat.
- **Scientific research** finds out more about the ecosystem, how it is threatened and how best to protect it.
- **Education** informs visitors about the taiga and why it should be protected.

Sustainable forestry

After trees are cut down there is replanting with native taiga species. Forestry plots are carefully managed to conserve key species.

Problems: parks and reserves

- **Migration**. Taiga species often migrate long distances. Unless parks and reserves are very large, they cannot protect migrating species.
- **Money**. Where taiga has oil and gas, governments face huge pressures to develop them. Exporting oil and gas can lift whole countries out of poverty.
- **Pollution**. The taiga is easily damaged by atmospheric pollution. However, if parks and reserves are far from cities, few tourists will visit them. Money from tourism helps parks fund their conservation.

Problems: sustainable forestry

Sustainable management is expensive and long term. It is usually only possible for large companies or when international organisations provide funding. In Russia, for example, most of the taiga forest is leased to hundreds of small-to-medium-sized companies for 25 or 50-year periods – less time than it takes new trees to grow. The companies are not interested in sustainable management; they want to maximise profits by clearing as much timber from the taiga as they can within the period of their lease.

Worked example

Russia's government is now promoting oil and gas industries in eastern Siberia. The population in eastern Siberia has declined by 16 per cent in recent years, and the government hopes that oil and gas will bring jobs to keep the region populated. The region is also close to Asian markets, bordering China. However, there are challenges. Winter temperatures reach –50°C. The nearest cities are 1200 km away. A pipeline from the oil fields to ports on Russia's east coast cost US$25 billion to build.

Read the newspaper article above about oil and gas developments in Russia's taiga area of eastern Siberia.

Identify **two** advantages of developing Russia's eastern taiga. **(2 marks)**

Jobs for local people.

Money for the government from exporting oil.

When you are asked to identify or name a number of things, you can use a bullet list or short statements like this.

Now try this

Using the newspaper article on this page, identify **two** disadvantages of exploiting the oil reserves of eastern Siberia. **(2 marks)**

Energy impacts

You need to know about three categories of energy resources (**non-renewable**, **renewable** and **recyclable**) and about the different ways that extracting energy can impact on the environment.

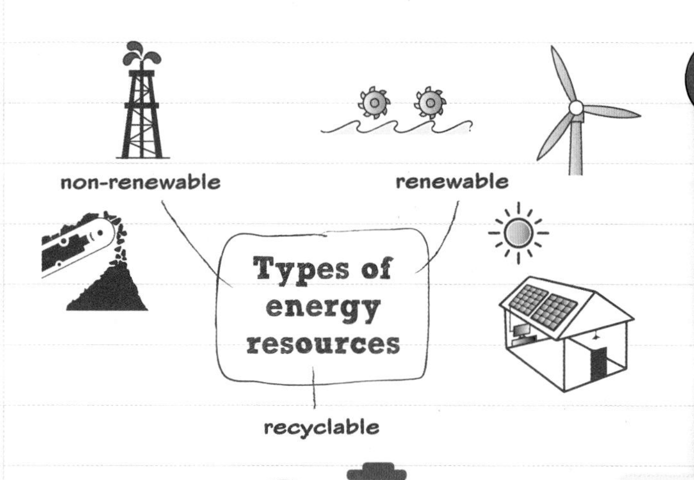

Worked example

The diagram shows different types of energy resource represented by icons.
Using your knowledge of energy resources, identify **one** type of non-renewable energy, **one** type of renewable energy and **one** type of recyclable energy. **(3 marks)**

Non-renewable – coal

Renewable – hydroelectric power

Recyclable – biofuel

Short answers are fine for this kind of question, but remember to use proper sentences when describing, explaining and assessing.

Deforestation and landscape scarring result from mining, especially open-cast mining

The highest carbon emissions result from the extraction and use of fossil fuels

Hydroelectric power dams flood valleys behind the dam wall

Wind turbines and solar panels spoil some people's enjoyment of landscapes

Now try this

Describe **two** possible negative impacts on the environment of developing recyclable energy resources, such as nuclear power or biofuels. **(2 marks)**

Had a look ☐ Nearly there ☐ Nailed it! ☐

Access to energy

The amount of energy that people use is not the same in each country of the world. Access to different types of energy is also not evenly distributed. You need to know the reasons why!

Energy use

Access to energy can depend on physical resources. For example, Iceland has the highest per capita energy consumption. Iceland has access to geothermal energy sources because it is on a plate boundary.

Per capita (per person) statistics can be tricky to deal with. Some of the countries showing high per capita scores on the energy use map are oil-producing countries with relatively low populations, such as Kazakhstan. That gives them a lot of energy per person, but doesn't mean they are necessarily also highly developed countries.

Worked example

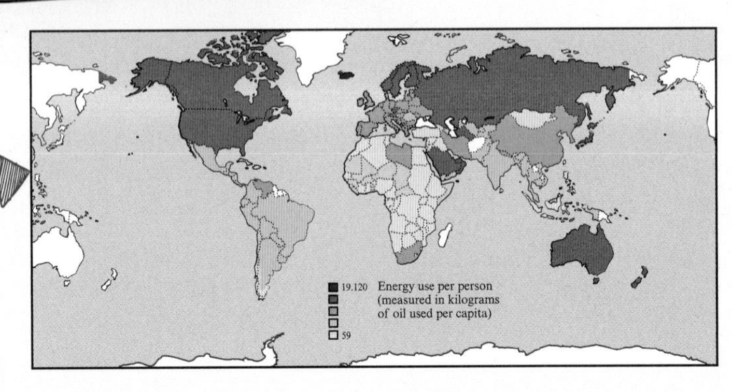

Referring to specific places on the energy use map, explain **one** factor causing variations in global energy use per capita. **(2 marks)**

Levels of development are one key factor. Countries that use the least energy per capita tend to be the least developed countries of the world. When people are very poor, they do not use much energy. They use traditional sources of fuel – wood or dung – for cooking and heating, not oil or gas. The most developed countries use a lot of energy because people use a lot of electricity in their homes, have central heating and/or air conditioning and drive cars.

Hydroelectric power

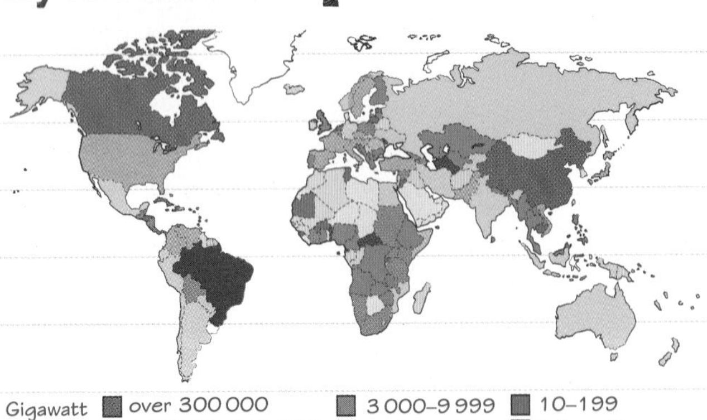

Not every country can generate electricity by hydroelectric power. Large volumes of water are needed, and steep drops in terrain (or massive dams).

Nuclear energy status

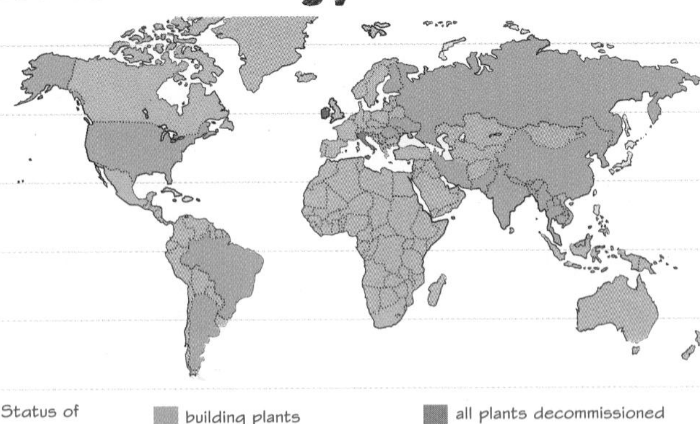

Access to technology can also affect access to energy resources. For example, nuclear power stations require highly sophisticated technology and expertise.

Now try this

Describe **three** ways in which climate could influence a country's ability to access renewable sources of energy. **(3 marks)**

Global demand for oil

The demand for oil is growing around the world due to economic development. However, really big oil supplies are only found in a few countries. Oil prices can be strongly affected by international politics affecting these countries, as well as economic changes.

Why is oil consumption rising?

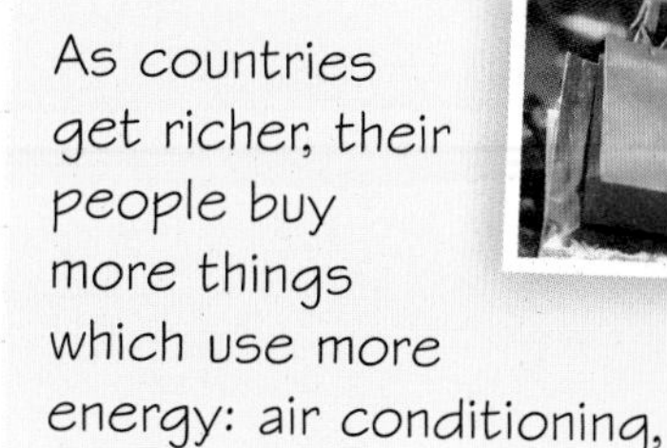

World population reached 7 billion in 2011. It is expected to peak at 9.5 billion by 2050. All these new people need energy too.

As countries get richer, their people buy more things which use more energy: air conditioning, cars, etc.

As new technology is developed, people want to buy new things or the latest version.

Price per barrel of US crude oil at end of each month

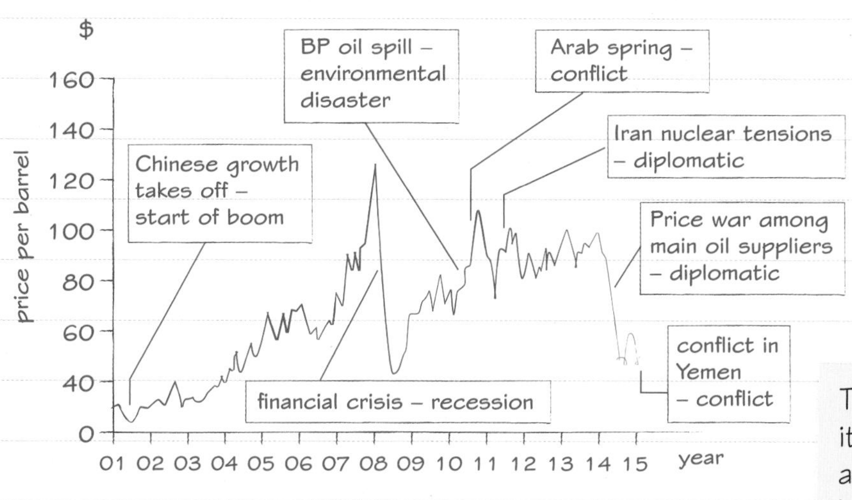

This graph shows that changes in oil prices are usually triggered by conflicts, or economic or political crises.

Oil consumption is measured in millions (m) of barrels per day.

This is a good answer because it uses the graph resource in a relevant way. It is also good because it sticks closely to what the question is asking.

Worked example

Using the graph of oil prices above, explain how China's development has affected global oil prices. **(4 marks)**

China does not have large oil reserves, but it has the world's fastest growing oil consumption: a 90 per cent change between 2000 and 2010, China's rapid industrialisation and economic development massively increased demand for oil in China causing the steep rise in oil prices between 2001 and 2008.

In 2008, however, the oil price fell very steeply. This was caused by the financial crisis ('credit crunch'). This financial crisis was global and meant that people around the world stopped buying so many things from China. China's industries stopped needing so much oil. When demand for oil is reduced, oil prices fall.

Now try this

The graph of oil prices shows that prices increased in 2010 after a huge oil spill near Florida and political turmoil in the Middle East. Suggest why these two events led to price increases. **(2 marks)**

New developments

Some oil and gas reserves are in remote, challenging areas or are stored in geologically complex ways. Sometimes it is worthwhile economically (or politically) to mine these challenging areas.

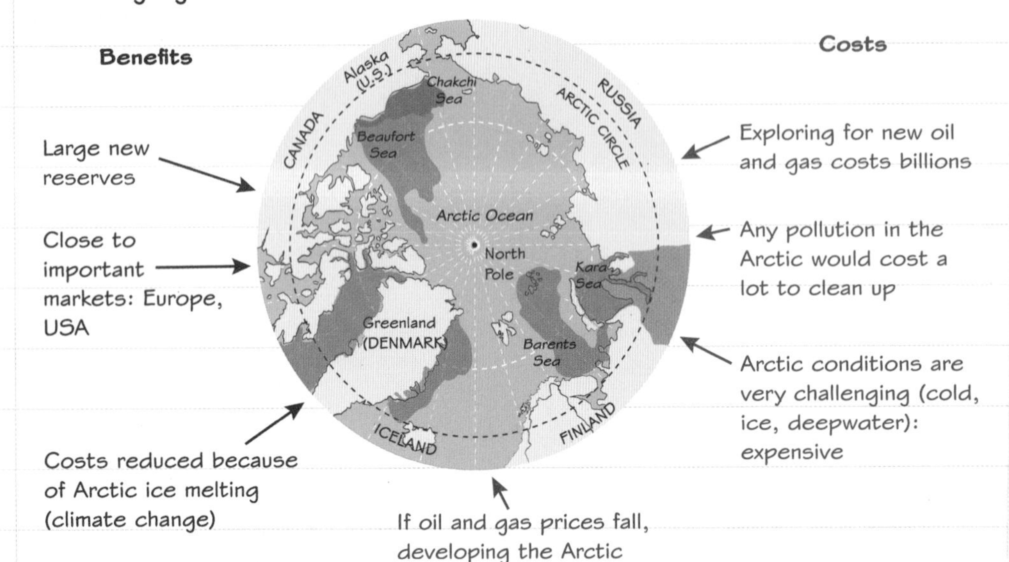

Rising global temperatures have reduced the economic cost of drilling for oil and gas in the Arctic. However, there are still significant challenges to developing new sources of oil and gas in this remote and ecologically sensitive area.

Environmental costs

Canada has the world's largest deposits of tar sand, a type of oil. Tar sand is difficult to extract, and toxic chemicals have to be used in the extraction process. Protesters are concerned that these chemicals are damaging people's health through air pollution, and threatening ecosystems (which are also at risk from oil spills).

Since 2011, the USA has increasingly used fracking to supply natural gas. This means it has used less coal for electricity generation, but some people are concerned about the impact of fracking on the environment, for example contaminating groundwater, causing subsidence and destroying natural habitats.

This answer is good because it uses specific detail that shows a good understanding of the processes involved in shale gas extraction that is clearly linked to the question.

Worked example

The graph opposite shows the results of a UK poll about whether the UK should go ahead with shale gas extraction (fracking) or not. Suggest one reason why people might oppose fracking in the UK. **(2 marks)**

People are worried that fracking for shale gas will contaminate groundwater because fracking injects chemicals into underground rock formations to help extract the gas.

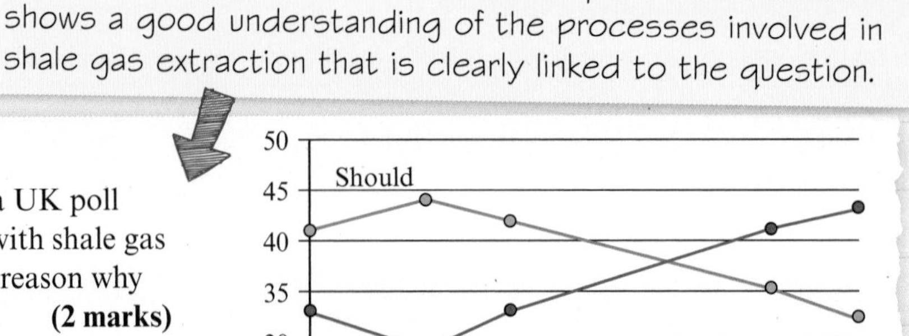

Now try this

Suggest **one** reason why some people think the UK should use fracking to extract shale gas. **(2 marks)**

Energy efficiency and conservation

Using energy more efficiently and reducing the amount of energy we waste will help our non-renewable energy supplies last longer, and will also reduce carbon emissions.

Energy conservation at home

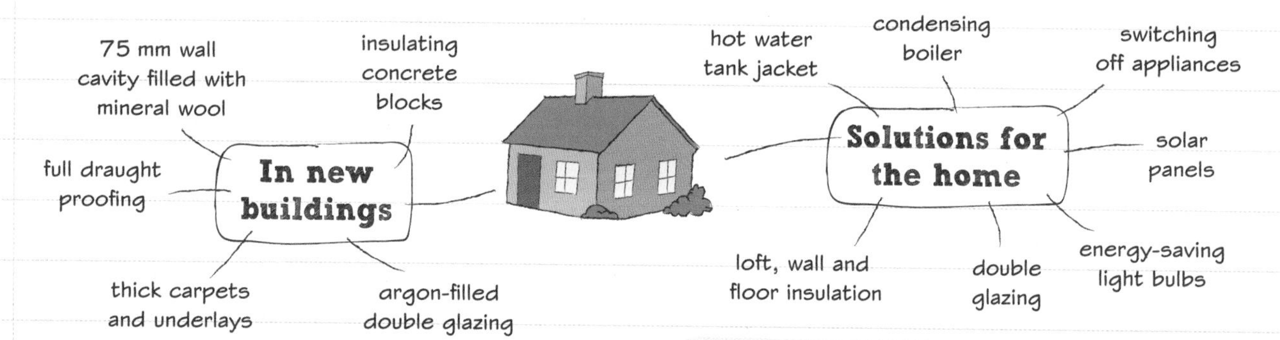

Worked example

Highlight the key words in the question to help you focus on what the question wants you to do.

Describe **two** ways to reduce domestic energy wastage. **(4 marks)**

People could reduce energy waste in their homes by switching off electrical devices rather than leaving them on standby. Experts estimate that 8 per cent of the energy used in houses goes on powering devices on standby. Also houses could be built with energy-efficient materials such as insulation blocks for walls to keep in heat, and the use of argon in double glazing would stop heat escaping through the glass.

UK transport energy consumption

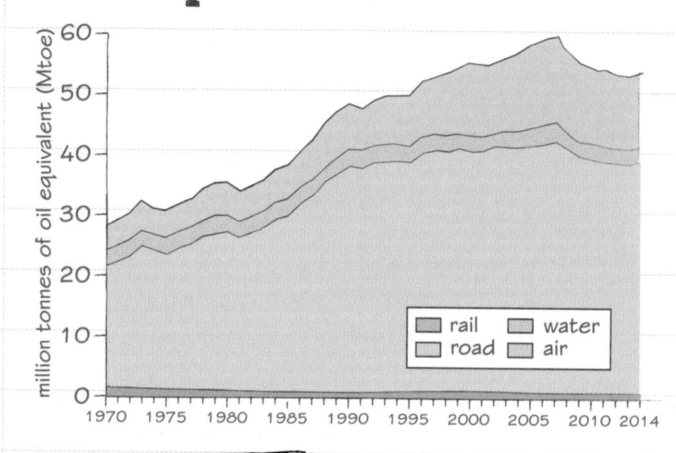

Now try this

Study the graph on this page. Between 1970 and 2014, road energy consumption increased from 21.4 Mtoe (million tonnes of oil equivalent) to 40 Mtoe. Calculate this increase as a percentage. **(1 mark)**

Transport and energy efficiency

Over three-quarters of the UK's oil use is for transport. Transport also contributes 22 per cent of the UK's carbon emissions. There are different ways to reduce the amount of energy used in transport in the future.

1. Changing transport use. Trains and buses are more efficient than individual cars because there is one engine to carry many passengers instead of many engines each carrying one passenger.
2. Improving engine efficiency. Decreases in energy consumption in the graph opposite are mainly due to new car engine designs that have lower fuel consumption than older models.
3. Improving energy conservation. Aeroplane design has focused on reducing drag, which reduces the amount of energy lost during flights.

Had a look ☐ Nearly there ☐ Nailed it! ☐

Alternative energy sources

Renewable energy sources (like solar, wind, HEP and biofuel) are alternatives to fossil fuels. Renewable energy can help countries reduce both their carbon footprints and their reliance on getting oil and gas.

Benefits of renewables

- low or no carbon emissions, lowers contribution to global warming
- inexhaustible, available for ever
- clean, no local air or water pollution
- widely available, one or more are likely to be available in most countries
- locally available, many can meet small-scale needs, especially useful in developing countries
- can reduce globalisation costs, e.g. CO_2 emissions, pollution, fuel, etc.

Costs of alternatives to fossil fuels

Cost of energy – for example, it costs more for a wind farm to generate the same amount of energy as a fossil fuel power station.

Geography – the best places for generating renewable energy are often a long way from the cities where energy is needed.

Extensive land use – wind farms, solar farms, hydroelectric power reservoirs and biofuel crops all take up a lot of land area. There may be conflict with how other people want to use the land – for example, for growing crops to feed people.

Impact on landscape – renewables are very visible and some people say they spoil the landscape; they may also create noise pollution (e.g. wind turbines).

Impact on local ecosystems – for example, deforestation to grow biofuel crops, birds being killed by wind turbines, valleys being flooded for hydroelectric power.

Worked example

Read the information opposite about hydrogen-powered cars. What are the costs and benefits of hydrogen as an alternative to fossil fuel? **(4 marks)**

Hydrogen and oxygen are available very cheaply everywhere. So energy security would be much better as countries would not have to rely on oil supplies from the Middle East.

Hydrogen fuel produces clean energy with no carbon emissions, reducing the impact of climate change.

However, hydrogen technology is very expensive to develop. This means that hydrogen cars are really too expensive for most people to buy. Another cost is infrastructure. Many more hydrogen fuel stations would need to be built all over the UK.

This car is powered by hydrogen. Hydrogen reacts with oxygen in a fuel cell to produce energy that charges an electric motor. The only exhaust is water vapour. This model costs £66 000 and can be driven for 300 miles on a full tank. There are fewer than 20 hydrogen fuel stations in the UK.

Now try this

Explain what is meant by energy security. **(2 marks)**

Attitudes to energy

There are contrasting views about energy consumption and whether people should reduce their carbon footprints or continue with current levels of energy consumption: 'business as usual'.

What is a carbon footprint?

It is a measurement of all the **greenhouse gases (GHGs)** individuals contribute to our environment as a result of our daily lives.

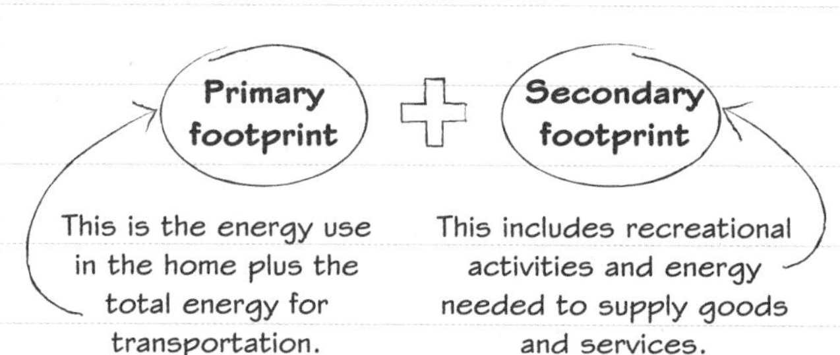

How is it measured?

A carbon footprint is written as kilograms (kg) of the **equivalent carbon dioxide** per person. The world average is 4000 kg, and the target to fight climate change is 2000 kg.

Changing attitudes

Climate scientists advise governments about the dangers of a 'business-as-usual' approach to energy consumption. Although environmental groups pressure governments to change energy policy, TNCs may resist changes as reducing energy consumption increases their costs.

Then again, if consumers are unwilling to pay more for renewable energy, governments will find it hard to make renewables a bigger part of the energy mix.

So what changes consumers' attitudes?

- **Education** – for example, government information about energy choices.
- **Environmental concerns** – for example, the impact of campaigns by environmental groups.
- **Rising affluence** – for example, can afford more energy-efficient options such as solar panels.

Worked example

As a country becomes more developed its carbon footprint tends to increase.
Explain reasons why this occurs. **(4 marks)**

Carbon footprints measure carbon emissions in different sectors, such as food, travel and housing. When people live in poverty, they have low carbon footprints because they travel by foot or bike, have small homes without heating and eat food produced locally. As countries become more developed, people travel by moped, have larger homes perhaps with air conditioning, and eat food that has been transported from far away. These mean big increases in carbon emissions in the different sectors that contribute to the carbon footprint.

Now try this

In 2015 the Science Museum in London announced that one of the world's leading oil TNCs would no longer be sponsoring the museum's climate change exhibition. Environmental campaigners were delighted. Explain why the TNC and the environmental campaigners might each have had different views about how climate change should be presented to the museum visitors. **(4 marks)**

Had a look ☐ Nearly there ☐ Nailed it! ☐

Paper 3 (i)

Paper 3 is People and Environment Issues – Making Geographical Decisions. Sections A and B have short-answer questions. Section C has 8-mark extended writing questions as well as short-answer questions. Section D has a 12-mark extended writing question (plus 4 marks for SPGST).

Section C is: **Consuming energy resources.**

As in Paper 2, the extended writing questions in Section C use resources – maps, diagrams, photos, tables, graphs – as part of the question.

Unlike in Paper 2, these resources are in the Resource Booklet. Be careful that you use the correct figure from the Resource Booklet to answer each question.

The central question for this Section C is: How can the growing demand for energy be met without serious environmental consequences?

From that come other issues, such as:

- reasons for/patterns of rising energy use
- costs and benefits of conventional and unconventional energy/alternatives
- different views on energy futures.

Assessment Objectives 3 and 4

In Paper 3, the 8-mark questions involve the command words 'Assess' or 'Evaluate'.

- 4 marks are for **Assessment Objective 3**
- 4 marks are for **Assessment Objective 4**

Assessment Objective 3 is about **applying your knowledge and understanding** to geographical issues, using evidence to come to a judgement.

Assessment Objective 4 is about **selecting the right skill or technique** to investigate the question and to communicate your answer.

What to aim for

These extended writing questions are marked by levels. A top answer will:

- **apply your understanding** to unpick the different factors involved in the question
- **put together a clear, logical argument**
- **use evidence** to decide which of these factors are more important than others
- **use your geographical skills** to get accurate information
- **use this accurate information** in all parts of your answer rather than just in one bit.

Now try this

Study the table below, which shows conflicting views about developing new unconventional shale gas sources (fracking) in the UK. Assess the reasons why some groups are in favour of the development of shale gas.

(8 marks)

UK government	'Shale gas has the potential to provide the UK with greater energy security, economic growth and jobs.'
Residents in fracking test area	'In the USA where they have fracking, water sources have been polluted by chemicals, and there has been subsidence and earthquakes.'
Environmental pressure groups (e.g. WWF)	'The landscape scarring and pollution threats of fracking are unacceptable, and it is wrong for the government to be promoting fracking when we need renewables to combat climate change.'
Fracking companies	'We are ethical companies with a green code of conduct. We take all necessary precautions to avoid any pollution.'
Local governments in fracking test area	'There are considerable benefits for local communities in shale gas development areas because a percentage of tax revenues from sales of shale gas will go back to the communities for them to spend on services.'

Paper 3 (ii)

The final question of Paper 3 Section D: (Making Geographical Decisions) is worth 12 marks (plus 4 for **SPGST**). You will use material from throughout the **Resource Booklet**, plus what you have learned from the rest of your geography course, to decide which of three options you think is the best. The Resource Booklet and the three options will be different each year. You will need to justify your choice in a piece of extended writing, which is where the 4 SPGST marks come in.

Examples of options

Don't worry that these options talk about content you are not familiar with – that will all be explained in the Resource Booklet, which is part of the exam paper.

Option 1: Drill for oil in the WAR
It would be possible to drill for oil in the Arctic coastal area of Alaska, in the area called the Western Arctic Reserve (WAR).

Option 2: Build the US Keystone XL pipeline
The USA could bring oil from the tar sands of Alberta, Canada, via a new pipeline to the USA.

Option 3: Develop more renewable sources of energy
The USA is working to develop even more renewable sources of energy such as solar, wind, geothermal and wave power so may not need as much oil in the future.

Worked example

Study the three options for the USA.

1 Select **one** option you think would be best for the environment and for the USA. Justify your choice.

- Use information from Components 1, 2 and 3 to support your answer.

(12 marks, plus 4 marks for SPGST)

I think the pipeline would be the best option. This is because a lot of people would get jobs and US states along the route would get more tax. It would also save the USA having to import so much oil.

I don't think that the environmental costs are a big problem. The Ogallala Aquifer does not appear to be a very important area for wildlife. The amount of carbon dioxide created might add to global warming but so far the changes to climate have been small.

The drilling for oil in the WAR is not a good option because the roads and drilling pads would pollute the wilderness environment ...

Try to be specific and say how many jobs or how much oil.

Here links could be made to previous knowledge. For example, how oil leaks from the pipeline could be prevented (by better monitoring sensors). Make sure your statements can be supported with evidence.

Make sure you discuss the advantages and disadvantages of your chosen option to help explain why you think this is the best option.

Remember that 4 marks are available for SPGST, so make sure that your spelling, punctuation and grammar are really good and concentrate on spelling specialist terms correctly.

Try to give specific examples to support your points, using information from the Resource Booklet.

Now try this

Good answers can be made for any one of the options that you'll see on the exam paper – there isn't one correct choice. Good answers also consider all three of the alternatives suggested by the options and then explain why the chosen option is the best.

Try practising this skill with these three options about revision strategies. Choose one that you think is the best and then justify your choice, referring to why it is better than the other two choices.

- **Option 1** Read a page of the textbook, then close the book and see what you can remember.
- **Option 2** Revise with a friend and test each other on each topic.
- **Option 3** Summarise key information on a card per topic, then memorise each card's contents.

Atlas and map skills

There are general geographical skills that you might need to draw on for any of the three exam papers. These include atlas and map skills. You will need to be able to describe the distribution and patterns shown on different kinds of maps.

Atlas maps

One of the most common types of map shows **distribution** – for example, the distribution of vegetation types, such as tropical rainforests.

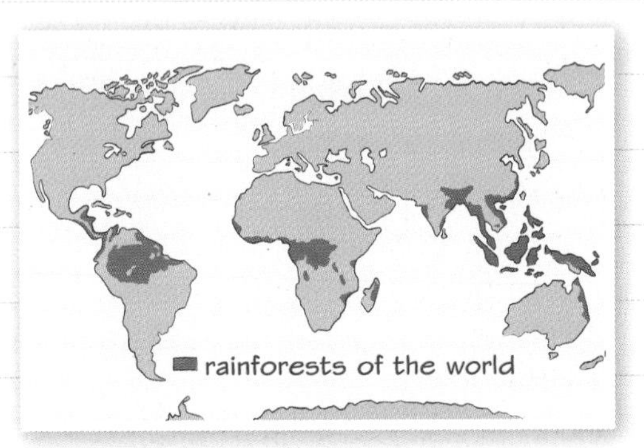

Atlases also contain maps which show:

- **climate zones** and global variations in **precipitation** and **rainfall**
- country boundaries (political maps)
- **height** and **shape** of the land (**relief**)
- population distribution (how people are spread within a region or country).

You may be required to use a combination of different types of information to answer a question.

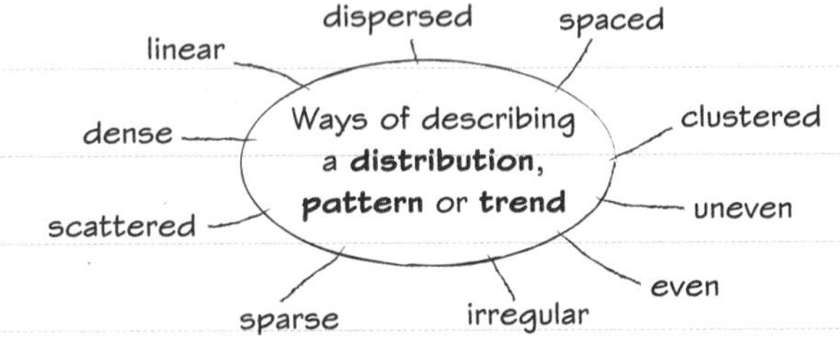

You can also use the letters GSE to help structure your descriptions of patterns:

G – General overall trend or pattern

S – Specific examples that illustrate the trend or pattern

E – Exception: note any anomalies that do not fit in with the general pattern or trend.

Don't fall into the trap of **explaining** why the pattern happens, unless asked to. Underline the command word in the question to help you focus your answer.

Worked example

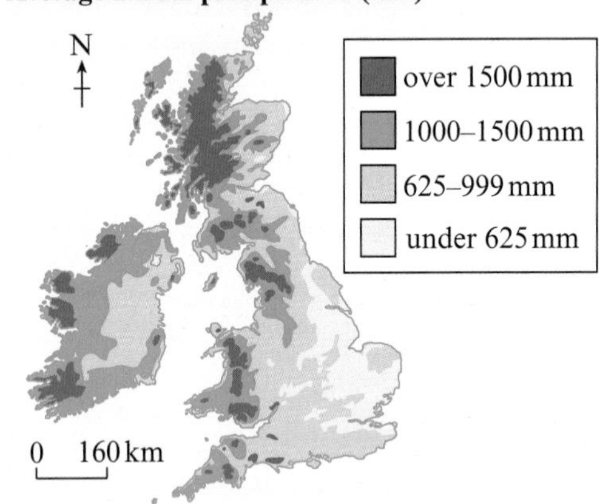

Analyse this figure, which shows the distribution of rainfall in the UK.

Describe the distribution of rainfall shown by the map. **(2 marks)**

Some parts of the UK, in the west and north, receive over 1500 mm of rainfall annually. The east of the UK is drier, with some parts of eastern England receiving less than 625 mm per year on average.

Now try this

Analyse the map above, which shows the global distribution of tropical rainforest. Suggest **two** reasons for the distribution shown by the map. **(4 marks)**

Types of map and scale

You need to be able to recognise and describe the distribution and patterns shown by the types of maps found in atlases, and deal with maps at a range of scales.

Satellite images and maps

Political maps which show the outline of countries

Maps which show the **distribution** of vegetation type, e.g. location of tropical forests

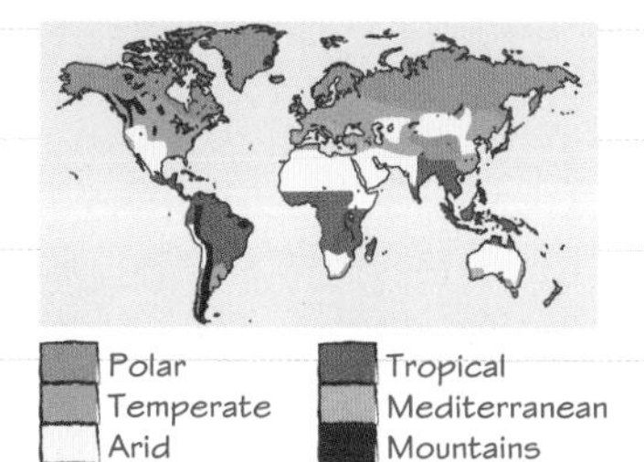

Climate zones which reflect global variations in precipitation and temperature

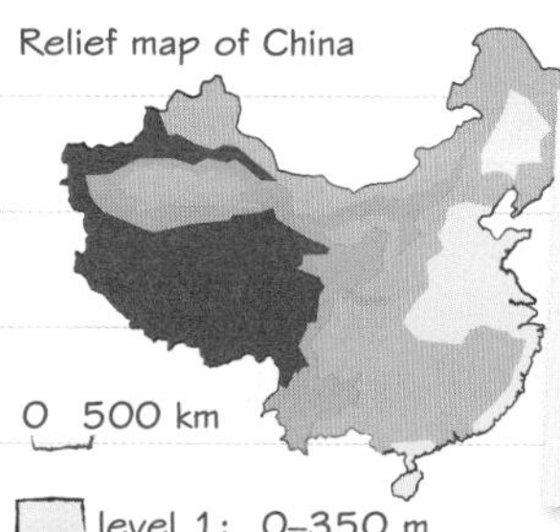

Relief maps showing the **height** and **shape** of the land

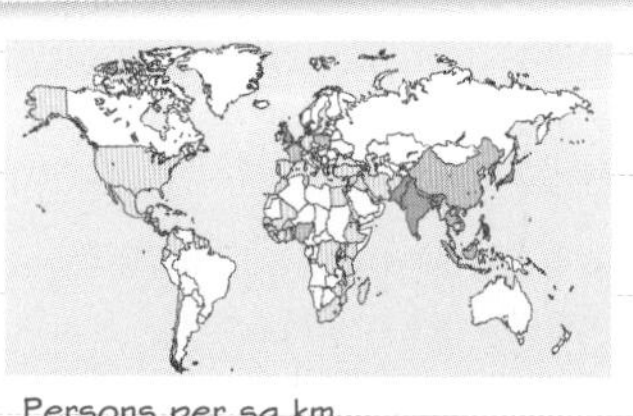

Population distribution maps which show how **spread** out people are within an area

What is scale?

A map's scale tells you how much smaller the area shown on the map is compared to the area in real life.

- For OS maps at 1:25 000 scale, 1 cm on the map represents 25 000 cm in real life (250 m).
- For OS maps at 1:50 000 scale, 1 cm on the map represents 50 000 cm in real life (500 m).

Large-scale maps show a smaller area in more detail.

Small-scale maps show a larger area in less detail.

Worked example

Study this 1:25 000 Ordnance Survey map extract of Bolton Abbey.

How far is it from the car park at Bolton Bridge along the B6160 to the car park at Bolton Abbey? **(1 mark)**

☐ **A** 0.5 km

☒ **B** 1 km

☐ **C** 10 km

☐ **D** 10.5 km

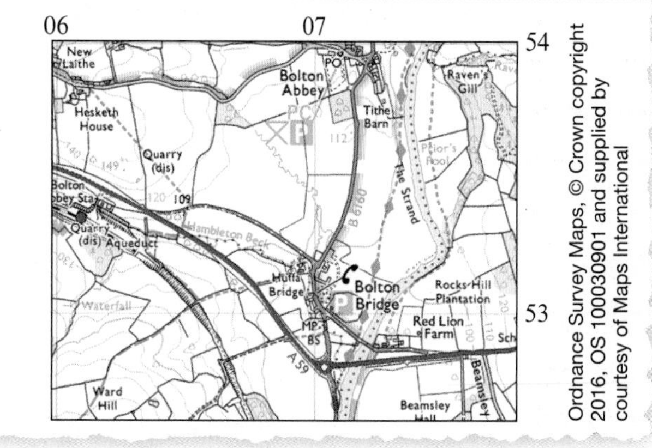

Now try this

The 1:25 000 map extract above of Bolton Abbey shows part of the River Wharfe in North Yorkshire. Which of the following is the best description of the River Wharfe? **(1 mark)**

☐ **A** A lowland river on a wide floodplain

☐ **B** A lowland river entering an estuary

☐ **C** A fast-flowing mountain stream

☐ **D** A river flowing through a gorge

Using and interpreting images

You should be able to respond to and interpret ground and aerial photographs and satellite images.

Different kinds of images

Ground-level photograph: shows lots of foreground detail. Use foreground and background to describe where things are in these types of photo.

Oblique aerial photograph: shows more of the area than a ground-level photo, and features are easier to identify than a vertical photo. But it is hard to judge scale for background features.

Satellite image: measures differences in energy radiated by different surfaces. False colour images convert this data into colours we can recognise. True colour images show us what the satellite sees, e.g. vegetation shows up red.

Vertical aerial photograph: these have a plan view, like maps. But details can be hard to identify.

The Five Ws

When working with photos, be sure to remember the Five Ws.

What does the photo show?

Why was it taken?

Who are the people in it (If there are people in the photo.)?

Where was it taken?

When was it taken (to indicate how long ago it was taken, what time of day, etc.)?

Worked example

Study this aerial photo of an open pit mine in South Dakota, USA. Identify **two** ways in which mining here has impacted on the landscape. **(2 marks)**

Landscape scarring, removal of forests

Read the question carefully for clues about what might be going on in the image. Link what you see in the image to the theories and processes you have studied.

Now try this

Describe **two** ways in which satellite photos like the one shown above can be used to monitor tropical cyclones. **(2 marks)**

Sketch maps and annotations

Drawing sketch maps

Sketch maps can be drawn using information from a map or photograph, or drawn in the field. They:

- ✓ show where basic features are located
- ✓ have simple labels
- ✓ are often drawn from an **aerial** viewpoint
- ✓ can be annotated to add more explanation or detailed information.

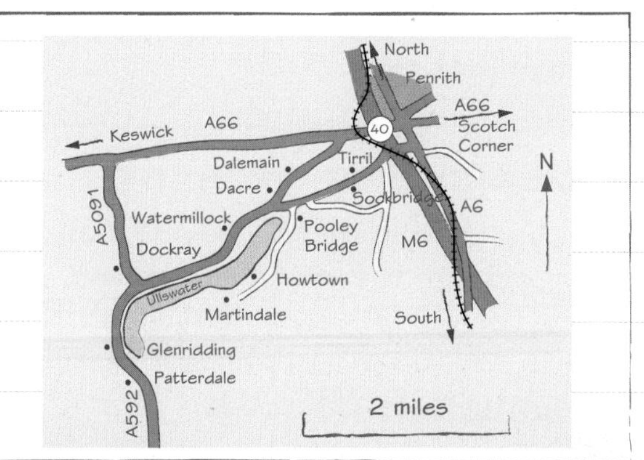

Sketching, labelling and annotating

- Photos and sketches are labelled and annotated in the same way.
- Only include the features that are relevant to the question.
- Draw clearly but don't worry about creating a work of art! Include a frame so that you can sketch within it.

This is a sketch of the photo above with labels and annotations

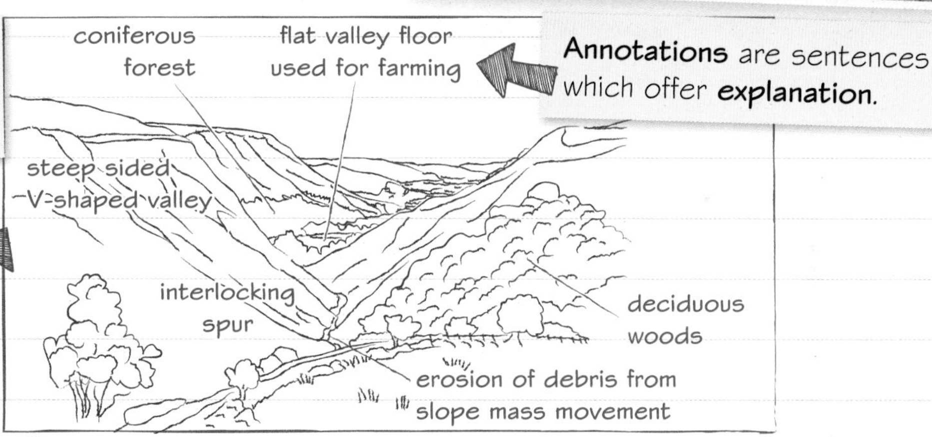

Annotations are sentences which offer **explanation**.

Labels are either one word or a short sentence which indicates what something is.

Worked example

Using an annotated diagram, explain how tsunami hazards can result from earthquakes under the sea. **(4 marks)**

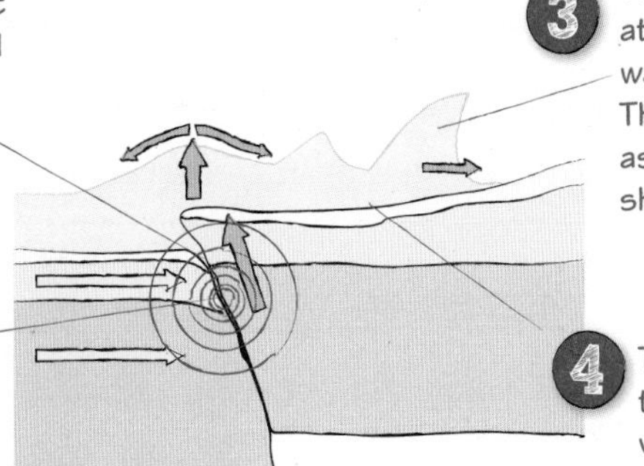

Now try this

Explain **one** advantage of using annotated diagrams to explain the impact of changes in river or coastal processes. **(2 marks)**

Physical and human patterns

You may be asked to use photos, maps or sketches to describe or explain **physical** and **human** patterns.

Describing patterns

You should learn to **describe** and/or **explain** the **distribution** and **pattern** of **physical features** (rivers and coastlines) and **human features** (settlements and roads).

Use the same technique you would use to describe any other type of pattern.

- ✓ You can use maps, photos and sketches to describe an area (e.g. rivers and coastlines).
- ✓ You can describe the site of a settlement using a map (e.g. settlements and roads).
- ✓ A photo or sketch can provide more detail about the **function** of the settlement.

For more about atlas and map skills, see page 106.

Worked example

Describe the section of the River Browney and its valley shown on the map extract. Use map evidence in your answer. **(3 marks)**

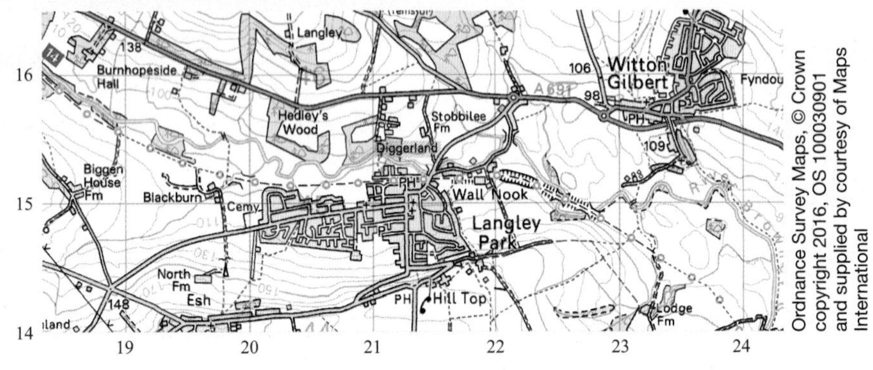

Make sure you use a good range of descriptive comments and map evidence.

This section of the River Browney shows the river moving from a V-shaped valley in the north-west of the map extract into a wider valley in the south-east. The V-shaped valley is only around 100 m wide at river level, with sides sloping up quite steeply to elevations of around 160 m. The river valley then widens to approximately 500 m, and is flat. There are a number of river meanders as the river valley widens, especially at grid reference 205153 and 233152.

Now try this

Suggest reasons why Settlement Y has grown. **(3 marks)**

Remember to focus on the **human** and **physical** reasons for the growth of the settlement.

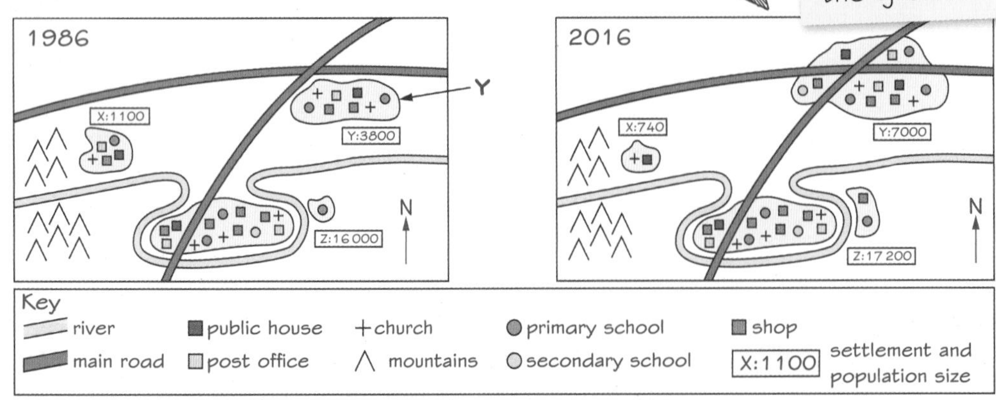

Land use and settlement shapes

OS maps show **land use**, **vegetation types**, **communications** and the **shape** of settlements. You may be asked to describe these features.

OS map features

On an OS map you will find information about:

- land use (settlements and farmland)
- vegetation (woods and parklands)
- communications (roads and railways).

You may need to describe the pattern or trend shown on the map.

The OS map key will help you identify different types of land use.

For more about atlas and map skills, see page 106.

Describing settlements

To **describe** and **identify** settlements remember the 3 Ss:

- **Site** – physical characteristics of the place
- **Situation** – location in relation to other places
- **Shape** – the way the settlement looks from an aerial view.

Remember also:
SAGA Slope (gentle or steep), Aspect (north, east, south or west-facing), Ground conditions (for example, floodplain), Altitude (height above sea level).

Settlement shapes

There are three types of settlement shape.

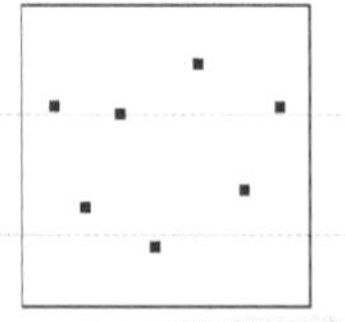

Dispersed

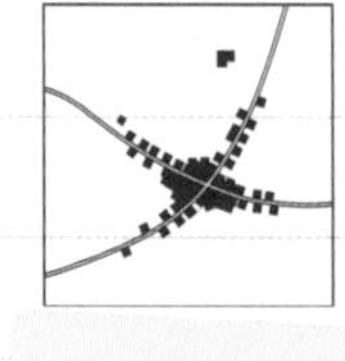

Nucleated

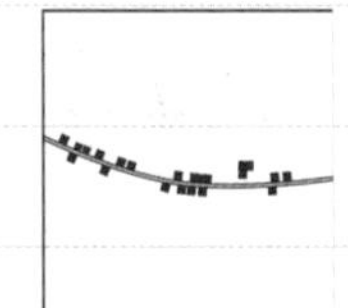

Linear

Worked example

Look at the OS map extract.

Find King's Caple in grid square 5628 and grid square 5629.

1 Put a cross in the correct box to describe the shape of King's Caple. **(1 mark)**

☒ **A** nucleated
☐ **B** scattered
☐ **C** dispersed
☐ **D** random

2 Explain your answer to **(1)**. **(2 marks)**

The buildings are grouped around one area; therefore it is nucleated.

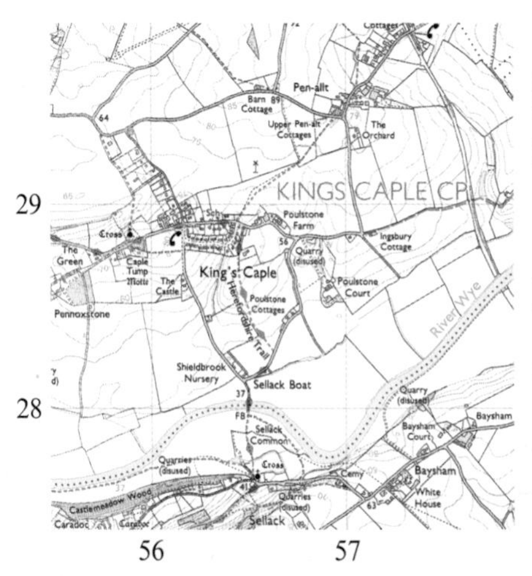

When describing patterns you will need to give map evidence to do well. Use grid references, road or settlement names, distances and directions.

Now try this

Look at the map extract in the Worked example.

Describe the distribution of woodland shown on the map. Use map evidence in your answer. **(3 marks)**

Human activity and OS maps

You need to be able to recognise different types of human activity on an OS map. You may be tested on these.

OS maps may show evidence of different types of human activity.

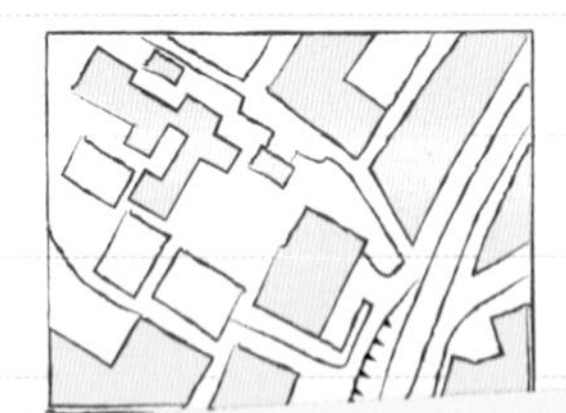

Industrial (e.g. factories and industrial estates)

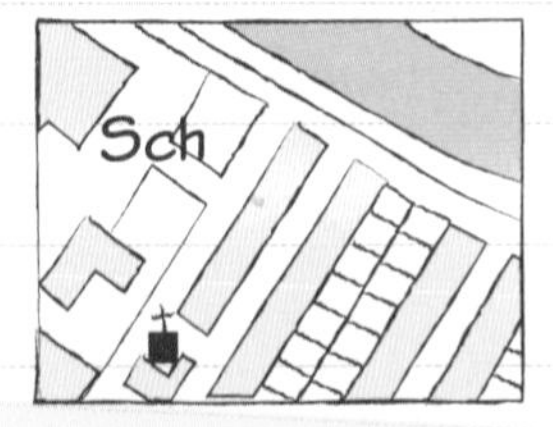

Residential (e.g. houses and flats)

Rural (e.g. forestry and agriculture)

Worked example

Look at the map extract of Ross-on-Wye.

Identify **two** pieces of map evidence that show non-residential activity by humans.

(2 marks)

At 620228 there is a farm and at 586233 there is an industrial works.

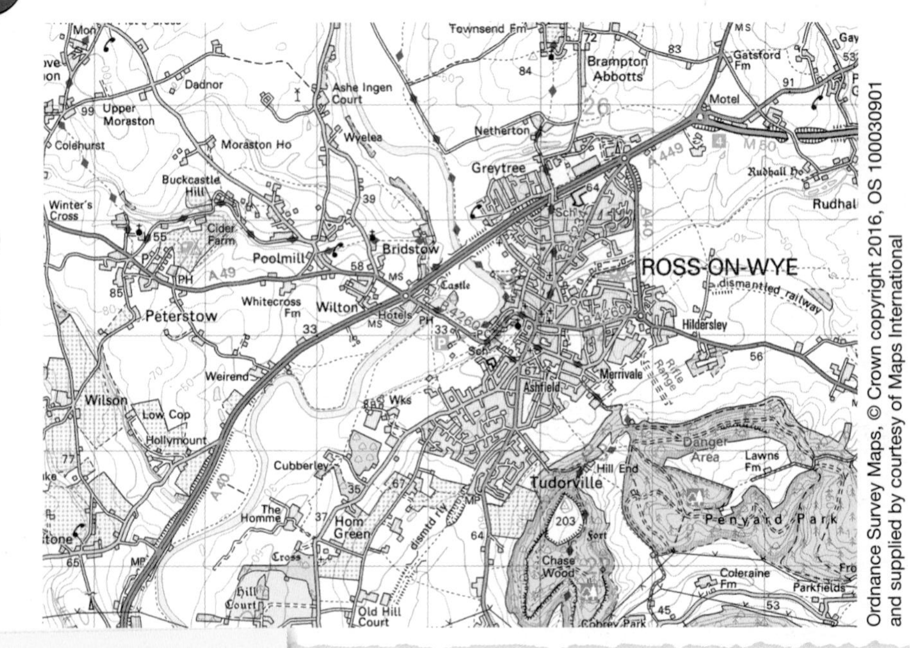

You need to name **two** different types of evidence you can see on the map. You can give a grid reference for each one to show that you know where they are.

Now try this

Ross-on-Wye has a population of around 10 000 people. Use the OS map extract above to complete the table with **two** more community facilities and your sketched version of the symbol for each. An example has already been provided for you. **(2 marks)**

Facility	Symbol
Car park	P

You need to include the name of the facility and the appropriate symbol.

Map symbols and direction

Accurate use of OS maps requires important skills, such as using the **key** to identify **symbols**, and use of the compass to work out aspect or direction.

Map symbols

OS maps use symbols to represent features.

- nature reserve
- place of worship with tower
- bridge
- public phone
- non-coniferous trees

Take care! When answering questions on direction make sure you start from the correct feature!

Compass points

You may be asked about the **direction** of features from each other.

You will need to learn the **eight** main compass points.

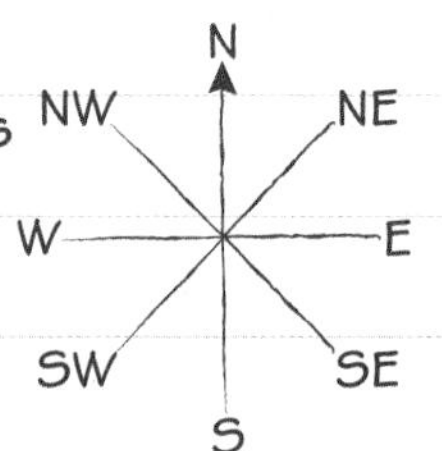

OS maps

This is a 1:50 000 extract from an OS map. This means that 1 cm on the map represents 50 000 cm (500 m) in real life.

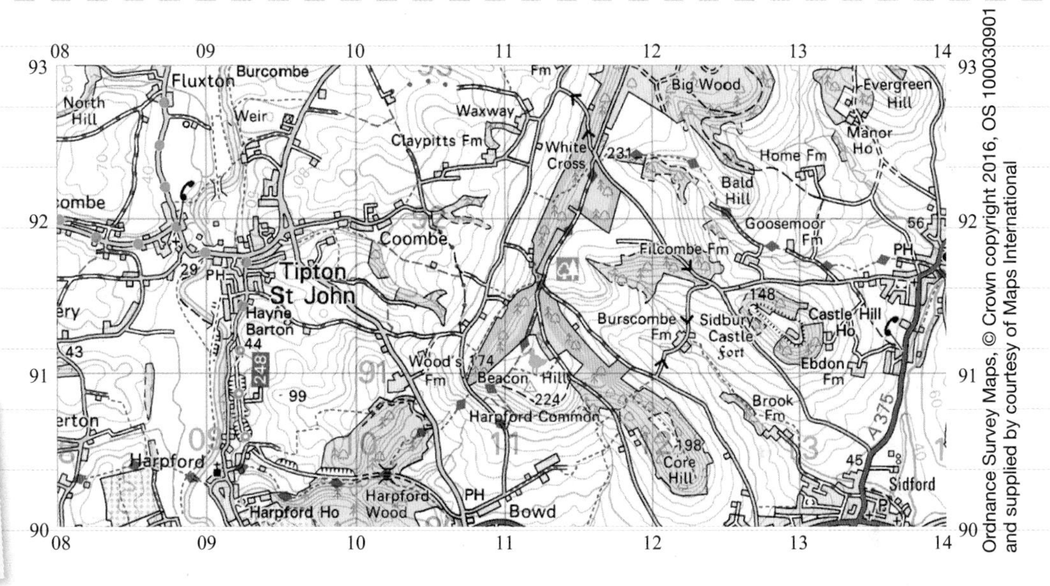

Look carefully at OS maps to spot the symbols and identify features.

Worked example

1 What do the following symbols mean on an OS map? **(3 marks)**

PC public convenience

coniferous trees

- - - - footpath

Make sure you know what the symbol means by using the key on the OS map. You do not need to memorise the symbols. Do not draw symbols too large and always use a **sharp pencil**.

2 In the box below, draw the symbol for an embankment. **(1 mark)**

Now try this

Look at the OS map extract above.

1 What symbol is found at 137913? **(1 mark)**

2 Which direction is Bald Hill in 1292 from Ebdon Farm in 1391? **(1 mark)**

Grid references and distances

Grid references are used to locate geographical features on an OS map. You need to know how to use grid references accurately.

Grid references

Each line on the map grid has a number. You can use these numbers to locate the features on a map.

You write the distance **along** (from the horizontal **easting** line) before the distance **up** (from the vertical **northing** line). For example, the shaded grid square has a 4-figure grid reference. To find this, you would go **along** the corridor, (13), then **up** the stairs, (02).

The telephone on this map has a 6-figure grid reference 138026.

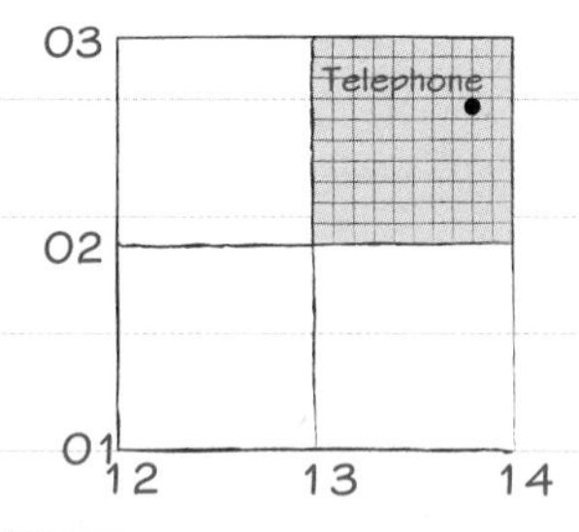

To write a 6-figure grid reference you have to mentally divide each grid square into 10 parts.

Measuring distance

You will need to work out **distances**. There are two types:

1. Distances from one point to another in a **linear** fashion are called **straight line** distances – sometimes called 'as the crow flies'.
2. Distances which follow a **curved** pattern, usually along a river or road, are called **winding** distances.

For the exam you will need a ruler to measure **straight line** distances and a 10-cm piece of string for **winding** distances.
The string should be provided for you if it is needed in the exam, but bring your own piece of string just in case.
Remember, you will be required to convert the distance measured using the ruler or string into kilometres (or metres if more appropriate) – use the scale line on the OS map to help you.

Worked example

Look at the OS map extract on page 113.

1 What is the name of the hill in grid square 1392? **(1 mark)**

Evergreen Hill

2 What is the 6-figure grid reference of the telephone in 1391? **(1 mark)**

137912

Watch out!
A pointer line from the telephone symbol (grid square 1391) shows the **actual** position of the telephone on the main road.

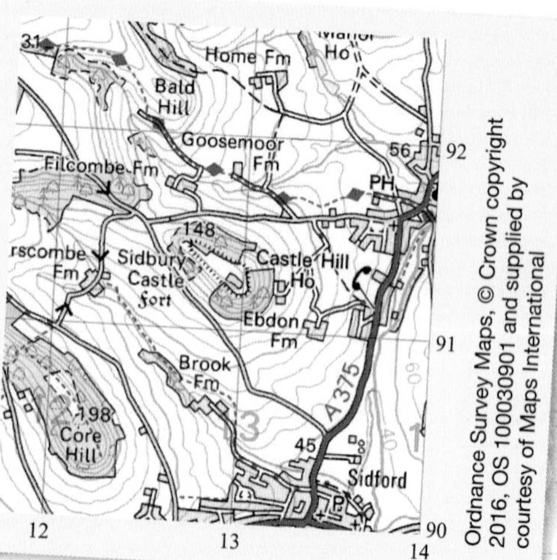

Now try this

Look at the OS map extract on page 113.

1 To the nearest kilometre, what is the straight line distance from Brook Farm in grid square 1290 to Home Farm in grid square 1392? **(1 mark)**

2 What type of woodland can be found in grid square 1192? **(1 mark)**

3 What is the 6-figure grid reference of the nature reserve near the centre of the map extract? **(1 mark)**

Cross sections and relief

A **cross section** is a visual representation of the landscape from an OS map. You may be asked to draw, label or annotate one, or comment on how you would complete it.

Drawing a cross section

1. Place a strip of paper along the given transect line.
2. Mark off the points where the major (brown) contour lines meet the transect line.
3. Mark the location of other features such as rivers, roads or high points.
4. Draw a line on the grid paper to be the x-axis of your cross section. Line the strip of paper up with this x-axis.
5. Mark off the height of each contour line using a neat cross. Join up the crosses with a ruler and a sharp pencil.

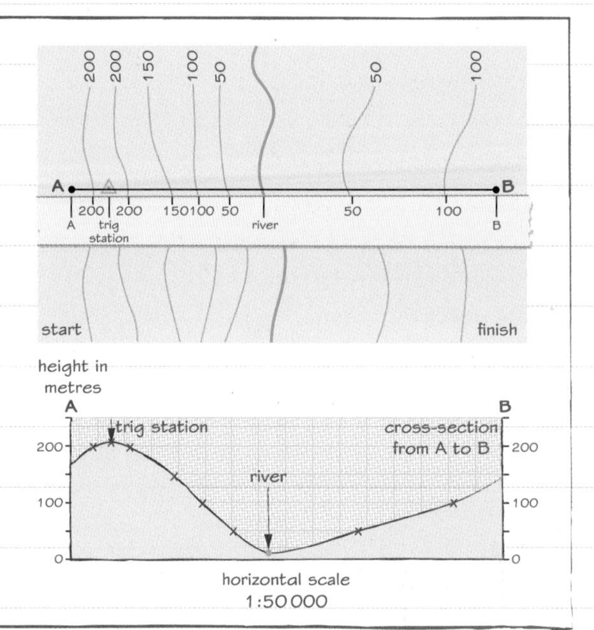

Slopes

The closer the contours, the steeper the slope!

There are different types of slope:

- **concave** slope
- **convex** slope

steep

gradual

Worked example

Look at the map extract on the right.
Put a cross in the box below to describe the shape of the land from A to B. **(1 mark)**

☐ **A** The land rises more gently towards the west.
☒ **B** The land at B is the highest.
☐ **C** The land at A is the highest.
☐ **D** The river goes through coniferous forest.

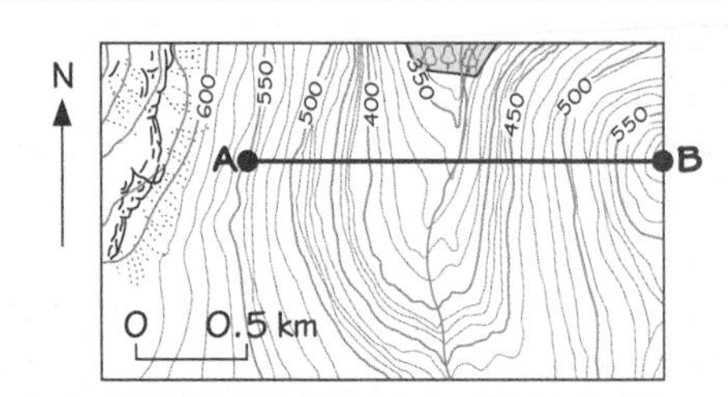

In multiple choice questions look at the number of marks as this is likely to tell you the number of answers required.

Now try this

Look at the OS extract in the Worked example.
Create a cross section of line A–B. **(4 marks)**

Graphical skills 1

In your exam you may be asked about different types of graphs and charts: to interpret the data they provide and also possibly about when it is appropriate to use a particular chart or graph.

Line chart

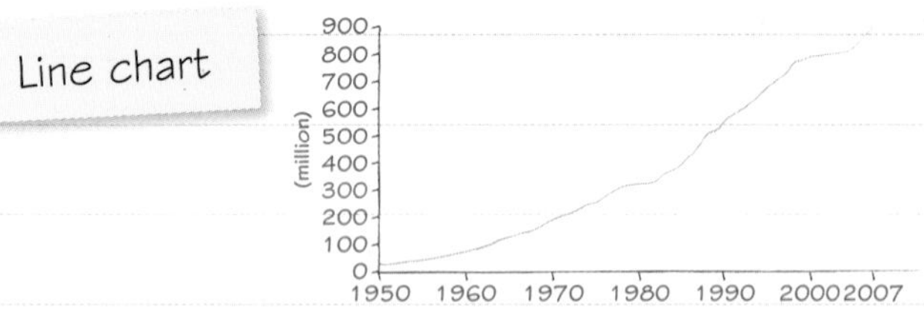

Line charts are used to plot continuous data. They are often used to show how something varies over time. Make sure you plot the points accurately and join the points with a continuous line.

Bar chart

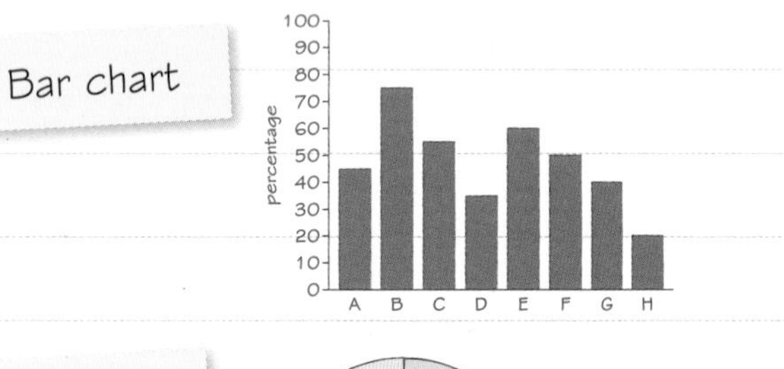

Bar charts are used to plot discontinuous data. Make sure you draw bar charts with a ruler to keep the lines straight.

Pie chart

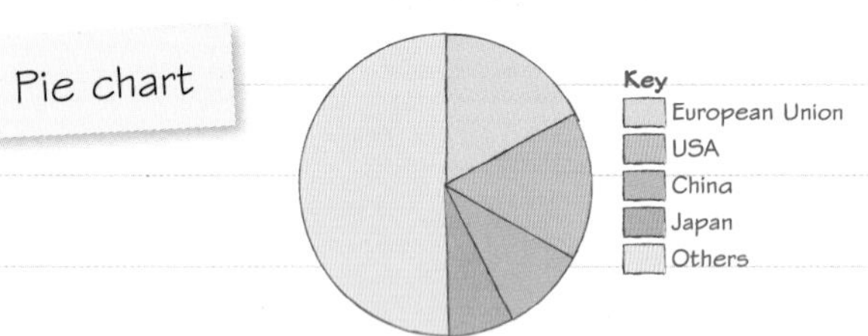

Pie charts show proportions. They are easy to read and fairly simple to put together. Data need to be converted into percentages first and then into proportions of 360° – the whole pie.

Scatter plot

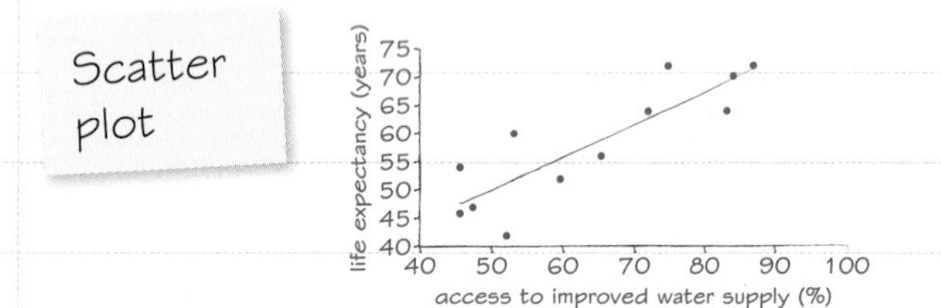

Scatter plots show the relationship between two sets of figures. It is the pattern the points make that is important, so don't join up the points. If the line of best fit slopes downwards it is a negative correlation; if it is upwards it is a positive correlation. Some scatter graphs do not show any relationship.

Pictogram

Pictograms represent data using appropriate symbols that are drawn to scale. They present data in a very clear way. However, detailed information can get lost. A key explains the relationship between the data and the pictogram.

Histogram

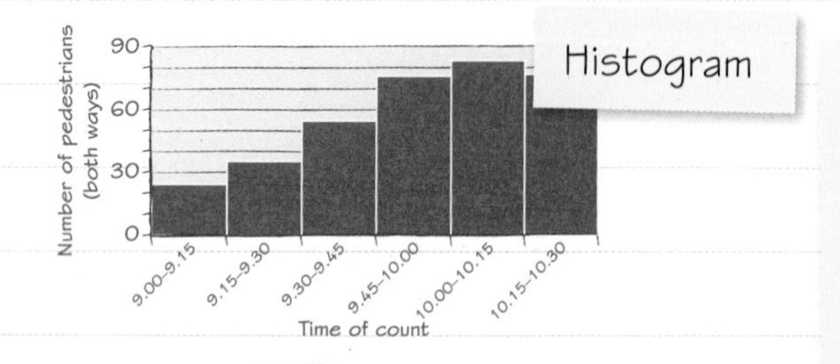

Histograms are used for continuous data. There are no gaps between the bars. The bars should be the same width for each category: this is called equal class intervals. The same colour is used for each bar because the data are continuous.

Now try this

What kind of chart or graph would you use to illustrate the following sets of data?

(a) Population growth in China from 1950 to 2010. **(1 mark)**

(b) The relationship between the size of settlements and the number of services in each. **(1 mark)**

(c) The proportion of people from different ethnic groups living in an inner city area. **(1 mark)**

Graphical skills 2

You should also know how to interpret population pyramids, choropleth maps and flow-line maps.

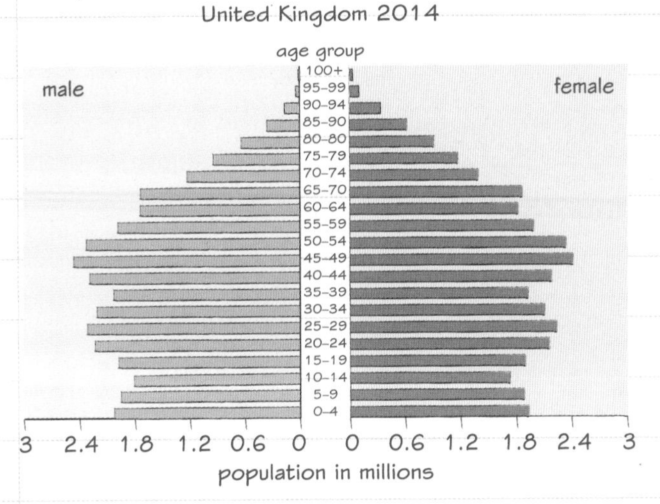

Population pyramids

Population pyramids are graphs that display data about the age structure of a population.

- The data are split into males and females.
- The data are presented in age groups.

This is how to interpret population pyramids:

- Developing countries often show a clear pyramid shapes: large numbers of children, lower life expectancy.
- Sides get straighter with development as birth rate reduces and life expectancy increases.
- Ageing populations start to look top-heavy.

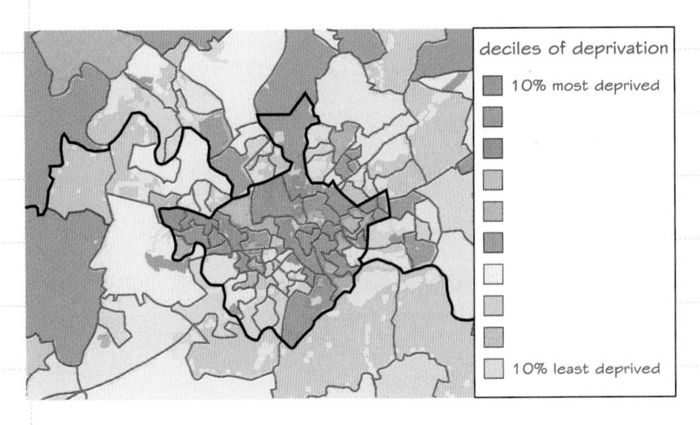

Choropleth maps

Choropleth maps are shaded so that each type of shade represents a particular range of values. The areas used for choropleth maps are usually ones that are used in lots of other ways, like government administrative areas.

Choropleth maps are very good for showing how something varies over a geographical area. One problem with them is that they can suggest abrupt changes between areas where actually changes are much more gradual.

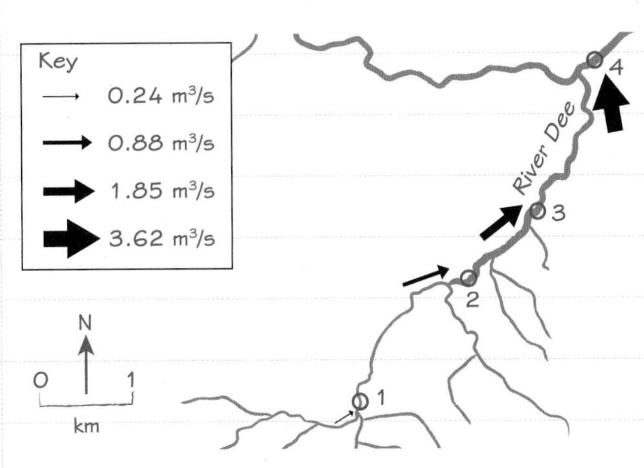

Flow-line maps

Flow-line maps are drawn so that arrows show the direction of flows, and the thickness of the arrows is proportional to the size of the flow.

Flow-line maps are easy to understand and give a clear indication of movement. The relative sizes of the flows can also be clearly seen. However, if lots of flows are going in the same direction the map can start to look very complicated.

Double-headed arrows can be used to show flows in two directions.

Now try this

Study the flow-line map on this page, which was used to present data from a geographical investigation on the River Dee. Suggest what was being measured at the four sites shown. **(1 mark)**

Numbers and statistics 1

You will use your mathematics and statistics skills in specific ways for particular topics in your fieldwork and you may also need them for any of the exam papers.

Proportion and ratio

Proportion – when two values are in direct proportion, then as one increases so does the other by the same percentage. If one decreases by the same percentage as the other increases, then this is called inverse proportion.

Ratio – indicates the relationship between two quantities, usually in terms of how many times one goes into another.

Equivalent ratios

You can find equivalent ratios by multiplying or dividing by the same number.

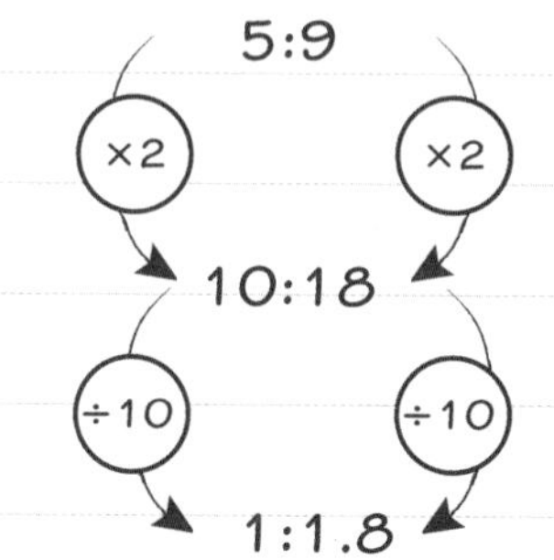

This equivalent ratio is for 1 :*n*. This is useful for calculations.

Percentage of an amount

To find the percentage of an amount:

1. Divide the percentage by 100.

2. Multiply by the amount.

For example, 84% of the UK's population of 64 million people live in England.

84 ÷ 100 = 0.84

0.84 × 64 = 53.76

So 53.76 million (53 760 000) people live in England.

One quantity as a percentage of another

To write one quantity as a percentage of another:

1. Divide the first quantity by the second quantity.
2. Multiply your answer by 100.

For example, the total energy production in a country is 32 million tonnes of oil equivalent. 8 million tonnes come from coal. To express 8 million as a percentage of 32 million:

8 ÷ 32 = 0.25

0.25 × 100 = 25

So 25% of the country's energy production comes from coal.

Finding percentage increase and decrease

To work out the percentage increase:

1. Work out the difference between the two numbers you are comparing:

new number – original number = increase

2. Then divide the increase by the original number and multiply the answer by 100.

To work out the percentage decrease is just the same, except this time the decrease is:

original number – new number = decrease.

Divide the decrease by the original number and multiply by 100.

Now try this

The population of Mumbai's metropolitan area has increased from 8 million in 1971 to 21 million in 2014. What is the percentage increase of Mumbai's population? **(1 mark)**

Numbers and statistics 2

Measures of central tendency help us to manage sets of data by giving us a way of describing them and comparing them easily. You need to know about the three main measures of central tendency – the median, mean and mode – as well as range, quartiles, interquartile range and modal class.

The **mode**: this is the value that occurs **most often**.

4 5 9 7 4 4
The mode of these six numbers is 4

The **mean**: to find the mean you add together all the numbers and then divide by how many numbers there are. Don't round your answer.

4 5 9 7 4 4
The mean of these numbers is 5.5
4 + 5 + 9 + 7 + 4 + 4 = 33
33 ÷ 6 = 5.5

The **median**: the median is the **middle value**. First write the values in order from smallest to largest. If there are two middle values, the median is halfway between them.

4 4 4 5 7 9
The median is 4.5

The **range**: the range is the largest value minus the smallest value.

4 5 9 7 4 4
The range of these numbers is 9 – 4 = 5

Modal class

When data are grouped, the modal class is the group that has the highest frequency.

Study the table opposite to see how this works.

Modal class = 21–25 mm

Size of stones (mm)	Number of stones
1–5	5
6–10	11
11–15	7
16–20	5
21–25	13
26–30	9
31–35	2

Quartiles and interquartile range

The **median** is the middle value: the halfway split in the data. Quartiles divide each half of the data into half: giving us quarters.

For the series of seven numbers 2 5 5 7 9 11 13, the **interquartile range** is the difference between the upper quartile and the lower quartile.

The lower quartile is (7 + 1) ÷ 4 = the 2nd value in the list = 5

The upper quartile is 3(7 + 1) ÷ 4 = the 6th value in the list = 11

Now try this

What is the interquartile range for the numbers: 2 5 5 7 9 11 13? **(1 mark)**

For the series of seven numbers 2 5 5 7 9 11 13, the **lower quartile** is the value that divides the lower half of the data into two halves. The **upper quartile** divides the upper half of the data.

ANSWERS

Where an example answer is given, this is not necessarily the only correct response. In most cases there is a range of responses that can gain full marks.

COMPONENT 1: GLOBAL GEOGRAPHICAL ISSUES

Hazardous Earth

1. Global circulation

The range is the difference between the lowest temperature (13 degrees) and the highest (35 degrees), so the answer is 22.

2. Natural climate change

C we are still in the Quaternary glaciation. We are currently in an interglacial within the ice age itself.

3. Humans and climate change

Coastal flooding from sea-level rise Melting ice caps and thermal expansion of seawater (water in the sea expands as it gets warmer, taking up more space) means that rising sea levels are a consequence of global warming. For communities living on low-lying coastlines (or low-lying islands), higher sea levels bring more chance of coastal flooding.

More frequent, stronger hurricanes are a consequence of global warming because hurricanes require warm water (26.5°C and higher) to form, and hurricanes are powered by heat energy – the warmer the water, the more intense a hurricane is likely to be. Increases in sea temperatures are therefore likely to produce this consequence.

Spread of pests and diseases The changing of climatic conditions within an ecosystem means that organisms from outside it that could not tolerate the conditions under which that ecosystem evolved are able to invade it as temperature changes. This can lead to a spread of diseases and pests that the original ecosystem inhabitants are not well adapted to resist.

Changes in farming Farming methods have to change as climate conditions alter – for example, dry farming methods may be required in regions that have previously relied on rainfall to give crops all the water they need. As different farming methods have differing levels of productivity, this could affect what types of food become available and how much of it farmers can produce.

Loss of glaciers Many of the world's most important rivers are fed by glacier meltwaters in spring – for example, the Colorado river. If glaciers shrink back or are lost completely, these rivers will supply much less water through the year and water supplies from them will become far less reliable in spring and through the summer.

In some parts of the world, warming atmospheric temperatures will **affect seasonal rainfall patterns, making climates drier**. This is particularly associated with rainfall brought by monsoon climate patterns in the tropics and subtropics. These are areas that rely on wet-season rainfall but are dry throughout the rest of the year so, if seasonal rainfall becomes less reliable, droughts can develop that continue for many years.

More floods are a consequence of global warming in some areas of the world, especially where the jet stream is involved in the development of low-pressure weather systems (such as in the UK). More frequent low-pressure systems picking up more moisture from warmer seas should mean more frequent rainfall and more intense rainfall, leading to more surface run-off and therefore a higher risk of flooding.

Biodiversity loss Your answer should include some of the following points:

- Many plants and animals have adapted to a specific temperature range. As temperatures increase, plants and animals will only survive if they can either move to areas that still have the temperatures and climate conditions they are adapted to, or evolve to adapt to the new conditions. The rate of warming is too fast for evolution and probably too fast for most plants to move to new areas (plants don't move themselves but rely on their seeds being carried to more suitable areas and germinating there).
- Increased temperatures also put stress on animal and plant species and these stresses make them more susceptible to disease and to pest attack.
- Increased temperatures may be more suitable for competitor animals and plants that move into new areas and outcompete the original species.
- Increased temperatures can trigger fires, which destroy large numbers of plants and animals.

4. Tropical cyclones

These months are when sea temperatures are at their warmest. Tropical cyclones depend on the energy from warm water (above 26.5°C) to form.

5. Tropical cyclone intensity

Category 5 Typhoon Haiyan had the strongest winds recorded for a tropical storm making landfall (coming onto land) – so strong that some people have suggested a new category might need to be added to the Saffir-Simpson scale.

6. Tropical cyclone hazards and impacts

Category 5 (Remember that the mathematical symbol > means greater than and the symbol < means less than). Tropical cyclones are low-pressure storms. The lower the pressure, the faster the wind that rushes in to replace the rapidly rising air.

7. Dealing with tropical cyclones

1.3 per cent (Remember that you calculate a percentage by dividing the first number by the second and multiplying the result by 100.)

8. Tropical cyclones

Questions like this one are looking for you to demonstrate your knowledge of geographical topics (in this case, ways in which developed and developing/emerging countries protect themselves against tropical cyclones) and to apply what you know, to do some analysis of the information and make a judgement at the end.

For this question you would be expected to agree that, generally speaking, developed countries like Japan and the USA are able to provide better protection than developing countries like Bangladesh, Vietnam, the Philippines or Myanmar/Burma because they can afford to spend more on protection. People in developed countries can usually afford cars to get them away from danger areas, and they have access to mobile phones, the internet and TV, which give them warnings about tropical cyclones and storm surges. In contrast, developing countries in particular may not be able to spend as much money on protection. Houses and infrastructure (roads, sanitation, health services) may be built of low-quality materials that are easily swept away by strong winds or storm surges. People may not have enough education to understand government warnings about tropical cyclones, or access to transport to get them away from the danger zone.

However, you could also consider the example of Hurricane Katrina, if you have studied it. Although the USA is the world's most developed country, the local government of New Orleans and the federal government of the USA had failed to spend enough money on keeping New Orleans protected from tropical cyclones – even a category 3 cyclone like Katrina. The poorest people in New Orleans were hardest hit by the disaster because they did not have cars in which to evacuate, or were worried that if they left their homes the police would not come to protect them from looters. Although 80 per cent of New Orleans was evacuated, the evacuation was chaotic, with people left stranded on the highway in traffic jams. For those who stayed, it was not clear where they should go for shelter, and there was not enough food and drink for people who had lost their homes.

The Hurricane Katrina example is therefore a good way to challenge the statement in the question and, perhaps, come to the following judgement: although in general the protection in developed countries is better, it is also possible for developed countries to be over-confident about their preparations, with disastrous results.

9. Tectonics

It used to be thought that the heat of the Earth's core was 'left over' from when the Earth formed – in other words, that the Earth is still cooling down from when the crust first started to solidify, 3.8 billion years ago. Now research suggests that around half the heat of the interior of the Earth is generated by radioactive decay – and also that it is this heat from radioactive decay (uranium and thorium) that rises up through the mantle to power the convection process. The other half of the Earth's heat, which *is* the heat left over from the Earth's formation, stays close to the core and is not involved in convection.

10. Plate boundaries and hotspots

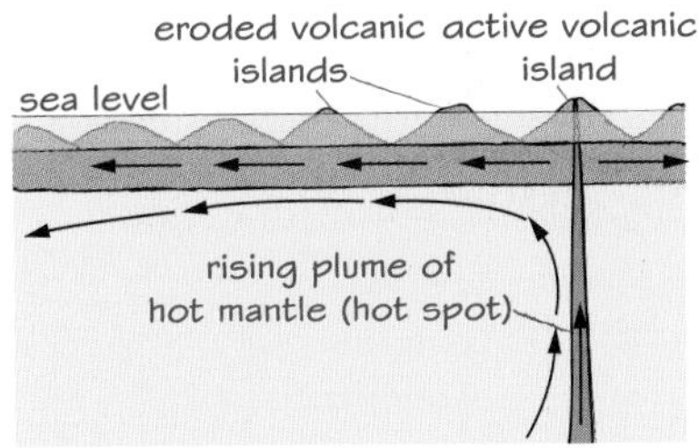

Volcanoes form over hotspots where magma is rising towards the surface due to mantle plumes. At the surface, the magma erupts through the thin crust. These volcanoes often rise above the ocean surface to form islands. Lanzarote formed as the crustal plate moved over the stationary hotspot.

11. Tectonic hazards

A tsunami is an ocean wave that is triggered by a number of different causes, including earthquakes under the ocean or near the ocean.
Two recent tsunamis are the 2004 Indian Ocean tsunami and the 2011 Tōhoku tsunami.
Tsunamis are a secondary effect of earthquakes that occur under the sea. (They can also happen as a result of earthquakes on land that trigger landslides under the sea.)

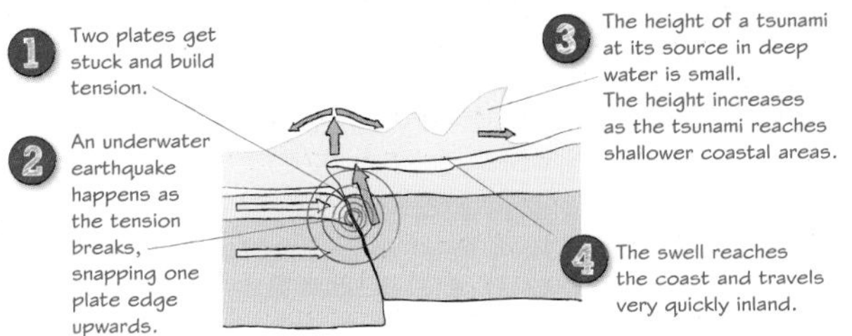

12. Impacts of earthquakes

Your answer will be based on your own located example information.

13. Impacts of volcanoes

Your answer will be based on your own located example information.

14. Managing earthquake hazards

The answers you give to this question will depend on the earthquakes you have studied for your developed country and developing or emerging country located examples.

15. Managing volcano hazards

Any one from: be ready to move out at short notice; practise evacuating the area; watch the volcano for signs of smoke and lava; be aware of any earthquakes occurring locally.

Development dynamics

16. What is development?

1 At least two of the following: economic measures show average figures only, so they do not show how wealth is distributed within a country; economic data are not always accurate and do not include non-official measures such as the cash economy, which is large in many developing and some emerging countries; economic measures do not give any indication of what that wealth is spent on; people consider other indicators – human, social and political indicators – as being just as important as wealth in measuring the development in a country or region.
2 GDP is just the Gross Domestic Product, whereas GDP per capita is the GDP divided by the total population. This gives a much clearer indicator – for example, China has a huge GDP but a much smaller GDP per capita because it earns a lot of money but there is a massive population.

17. Development differences

Answers could include the following:
A very youthful population can lead to social unrest if economic problems mean that a large number are unemployed.
An ageing population with long life expectancy can lead to healthcare problems because there are a lot of old people with complex health needs, while the number of younger people working and paying tax to fund healthcare is reducing.
Countries with high fertility rates need to invest in healthcare for mothers and babies, and will need to build more schools for the growing numbers of children.

18. Theories of development

Any two from:
1 It is out of date; it is based on the development of European countries when conditions were very different in the 18th and 19th centuries; it assumes all countries start from the same place; it does not take into account any variables of a country's resources, population or climate hazards; it only sees a continuous growth cycle – some countries have stalled at a certain stage.
2 Dependency theory says that richer states exploit poorer ones – poorer states do not benefit from capitalism because the rules are set by the most developed countries, which can set trade restrictions, and loan money with high interest rates and conditions attached, etc.

19. Types of development

Answers could include one of the following.
Governments contribute to globalisation by:
- setting up free trade areas that encourage TNCs to locate there
- agreeing low tax arrangements for TNCs, because this encourages TNCs to set up operations in different countries

- establishing trade agreements that encourage or discourage exports from developing and emerging countries
- passing laws that make global transfers of money and investment easier (encouraging globalisation) or putting more controls on these transfers (discouraging globalisation)
- making countries politically stable, which encourages TNCs to invest there
- encouraging free trade policies – for example, removing trade barriers designed to protect a country's industries.

20. Approaches to development

Your disadvantages of TNC investment could include:

- TNCs pulling out of a country if they can make more money by locating elsewhere
- poor working conditions compared to conditions in similar plants in developed countries
- although TNCs may locate production in a developing or emerging country, they often locate higher-paid, higher-skills roles (e.g. marketing, research and development) in developed countries, meaning that developing countries remain stuck as resource extractors and producers
- while TNCs themselves may insist on good health and safety regulations, they may outsource a lot of production to local suppliers who compete to give the lowest price to the TNC in order to get the contract, and then exploit their workers through low pay and poor safety conditions.

21. Location and context

A good place to look for information for a fact file on your case study country (apart from in your own notes from class) is the CIA's *World Factbook* – look it up online.

22. Globalisation and change

Key causes of demographic change in your case study emerging country are likely to be changes to fertility and death rates and rural–urban migration. Key consequences are likely to be the creation of inequality (between core and periphery, between younger and older people, between men and women), demographic consequences such as a decline in the birth rate in urban areas, and environmental consequences such as pollution.

23. Economic development

For your table you are aiming at three positive and three negative impacts of economic development and globalisation on people and on the environment. Check your notes from class to see what you can add to the information in this book.

24. International relationships

To find your case study country's top three trading partners, try looking online at the Observatory of Economic Complexity. This has some excellent data visualisations for exports and imports for a wide range of countries.

25. Costs and benefits

Make sure you have all the information you need for your emerging country case study.

Challenges of an urbanising world

26. Urbanisation trends

11 per cent.

27. Megacities

16.3 per cent. To calculate a percentage increase, first work out the difference between the two values – in this case 8.6 million and 10 million, so 1.4 million. Then divide 1.4 by 8.6 and multiply the result by 100 to get your percentage increase: 16.27, which rounded up to one decimal place is 16.3 per cent.

28. Urbanisation processes

Reasons for decline: economic changes due to there being fewer jobs in London **(1)** (you might have put damage from the Second World War, but that actually wasn't a major factor). Heavy industry in London declined, jobs moved out to the suburbs and London's port became less important.
Reasons for growth: economic changes that meant there was a lot more work in London **(1)**. The main driver of this growth was the boom in London's financial services. Redevelopment of areas like London's East End (the old dock areas) was important too.

29. Differing urban economies

Answers could include (1 mark for anything similar):

- new arrivals (migrants) to the city are able to support themselves
- doesn't require skills / education qualifications which means rural migrants are not excluded
- informal workers may learn skills that they can later use in formal employment
- easy to start – no need for premises, and rents and overheads are low
- flexible – if one member of a family has started an informal business, other family members can join.

30. Changing cities

Answer **D** is the best explanation as planning regulations would include laws on what type of industries are allowed to locate near residential areas. Accessibility would relate more to the transport links of the area, so are not so relevant in this example. Availability could be a factor – if that was the only land available. However, it is not the best explanation as the question makes a point about being 'away from residential areas'. Cost could be a factor too, but again it is not the best explanation as the question makes no mention of the land involved being cheaper than elsewhere.)

31. Location and structure

You need to know about the structure of your megacity, including the CBD, inner city, suburbs, urban–rural fringe.

32. Megacity growth

You need to know the reasons for past and present trends in population growth (rates of natural increase, national and international migration, economic investment and growth) for your megacity.

33. Megacity challenges

You need to know the challenges for people living in the megacity caused by rapid population growth, including problems with housing, with water supply and waste disposal, with poor employment conditions, and limited service provision and traffic congestion.

34. Megacity living

You should have located a wealthy area and a poor area in your megacity, and have identified at least one reason to explain the location of each area.

35. Megacity management

You need to know about the advantages and disadvantages of city government (top-down) strategies (managing water supply, waste disposal, transport and air quality) and you also need to know about the advantages and disadvantages of community and NGO-led bottom-up strategies (city housing, health and education services).

Extended writing questions

36. Paper 1

There are different ways you could answer this question and there is no one 'right' judgement to come to – although you should aim to assess at least two points about natural causes and two points about human effects on current greenhouse warming. Your answer will probably include some of the following points.

For **Assessment Objective 2** you should show your understanding of causes of past climate change and current global warming.

Points you might include for Assessment Objective 2
For past climate change (both warming and cooling):
- orbital changes (different types) and how these occur on very long timescales
- sunspot cycles (11 year cycle)
- volcanic eruptions (cooling).

For current global warming:
- rise in greenhouse gases concentrations in the atmosphere
- greenhouse effect as a natural process
- role of human activity in rapid increase in atmospheric greenhouse gases
- types of human activity that are producing the 'enhanced greenhouse effect'.

For **Assessment Objective 3** you should use your understanding to consider both sides of the argument – so you need to look at whether current global warming has different causes from the causes of past climate change or not.

Points you might include for Assessment Objective 3
Arguments for it having a different cause:
- Human activity is the cause of the enhanced greenhouse effect and human activity was not involved in past climate change.
- The enhanced greenhouse effect is something humans could choose to control by changing activities to reduce emissions – we cannot do anything to control orbital changes or other natural causes.
- The enhanced greenhouse effect is a rapid change, leading to rapid changes in climate – past climate change has happened very slowly or, in the case of volcanoes, quickly but temporarily.

Arguments for it having a similar cause:
- Ice cores show that increased atmospheric CO_2 has been associated with past climate change, so there could be a natural component to current global warming.
- Natural processes such as sunspot cycles, the El Nino effect and orbital cycles have not ended – they could also be contributing to current global warming – so although human activity may be a major cause, it is enhancing a natural process.

COMPONENT 2: UK GEOGRAPHICAL ISSUES

The UK's evolving physical landscape

37. Uplands and lowlands

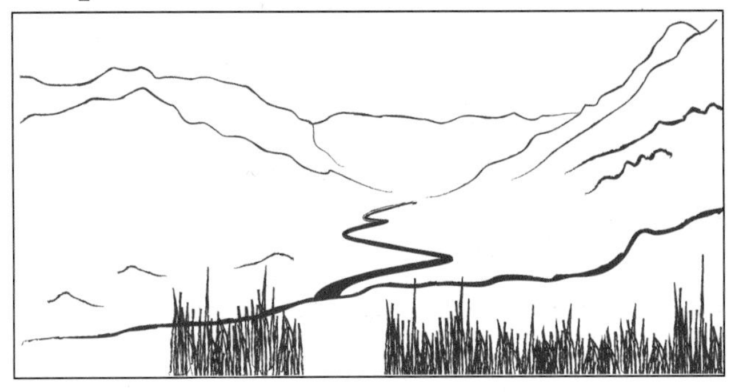

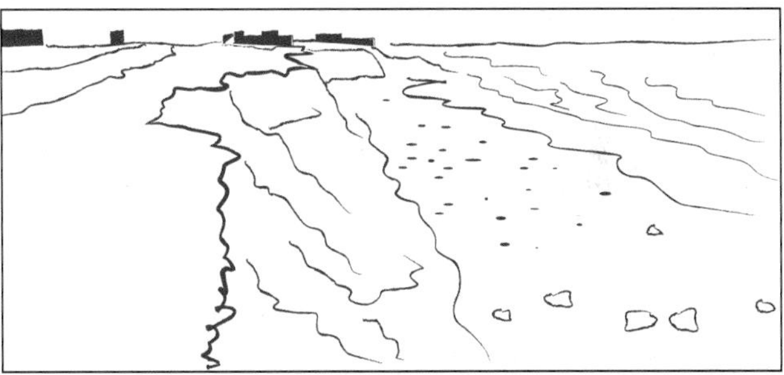

38. Main UK rock types
granite

39. Physical processes
A

40. Human activity
The river meander acted as a natural moat to protect the original settlement from attackers.

Coastal change and conflict

41. Geology of coasts
headlands and bays.

42. Landscapes of erosion
The sea erodes the base of the cliff forming a notch. Over time, the notch gets bigger and the cliff is undercut and eventually collapses. Debris from the collapsed cliff is washed away by the sea, exposing a wave-cut platform underneath.

43. Waves and climate
Any two from:
- Hydraulic action – the force of the wave hitting the rock often forces pockets of air into cracks in the cliff. This helps to break up the rock.
- Abrasion – the waves pick up stones and hurl these against the cliff and this wears away the rock.
- Attrition – the pebbles carried by the waves themselves get rounder and smaller as they are hurled against the cliffs and bash against each other.
- Solution – seawater dissolving rocks.

44. Sub-aerial processes

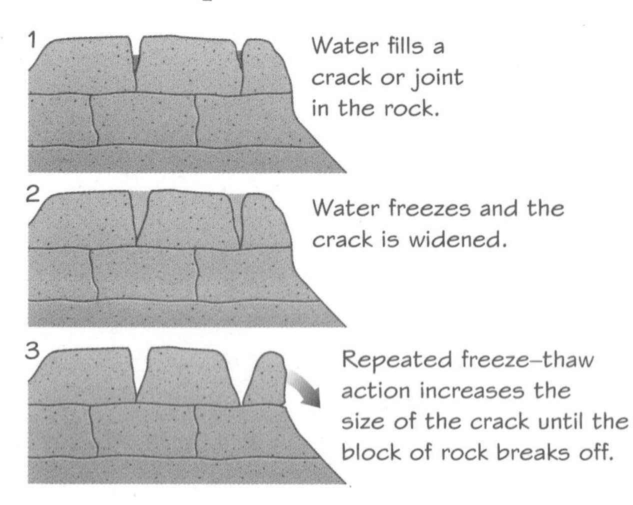

45. Transportation and deposition

When waves approach the coast at an angle, the swash as they break pushes sand and gravel up the beach at the same angle. The backwash drags the sand and gravel back down the beach at a 90° angle to the coastline due to gravity. This gives a zigzag movement of material along the beach and this repeated process is called longshore drift. The smallest material is sand and this is moved easily so it ends up at the top of the beach. The pebbles are heavier and are not moved as far.

46. Landscapes of deposition

Sand and pebbles are moved along the coast by longshore drift. At a bend in the coast the drift weakens away from the coastline and deposits material, and over time sediment builds up. If it builds up into a feature that goes right across the mouth of a bay, it is called a bar.

47. Human impact on coasts

Answers could include:

- Agriculture can increase soil erosion and sedimentation, increasing the risk of mass movement and coastal erosion.
- Draining coastal marshland to make it into farmland creates a very different coastal landscape.
- Increased amounts of sediment can be washed into streams and rivers because of agriculture, which may affect coastal deposition
- Trees and bushes are sometimes introduced by farmers to act as windbreaks near the coast: these can help stabilise sand and change beach profiles.
- Agriculture can have a positive impact by creating and preserving wildlife habitats.

48. Holderness coast

The Holderness coast is on the east coast of the UK. The geology there contributes to the rapid rates of coastal erosion because the rock type is soft boulder clay, which is easily eroded and prone to slumping when saturated. This coastline is also exposed to strong waves from the North Sea, which gives the waves a lot of energy to attack the coastline.

49. Coastal flooding

Climate change is likely to increase the risk of coastal flooding in the UK for two main reasons. Sea levels will rise as the sea temperature increases, putting low-lying land on the coast at increased risk of flooding or even total submersion. A rise in air temperatures will increase the frequency and strength of storms, which will increase the eroding power of waves as well as increasing weathering and mass movement of land on the coast.

50. Coastal management

Groynes are an example of a hard engineering method of coastal management. A benefit is that they reduce coastal transportation by preventing the sea moving sand away from the coastline, which helps protect the coast from wave erosion. A cost is that preventing the sea from transporting material at one location exposes other areas of the coastline to increased erosion.

Fieldwork: coasts

51. Investigating coasts: developing enquiry questions

The answer to this question will depend on your own fieldwork location. Being able to compare sites is often a key advantage. Either the coastal processes are not too different, which allows you to compare different management approaches, or the management approaches are similar, which allows you to compare differing coastal processes.

52. Investigating coasts: techniques and methods

The image of measuring a beach gradient with a clinometer is a quantitative fieldwork method. It generates numerical data. Collecting people's views by questionnaire is a qualitative method. It generates information that is not numerical but helps with understanding the geographical issue.

53. Investigating coasts: working with data

A

River processes and pressures

54. River systems

In the Afon Nant Peris river in North Wales the gradient is very steep in the upper course **(1)** (or your answer could say gently sloping in the middle course, or gently sloping in the lower course). The discharge is small in the upper course **(1)** (or your answer could say large in the middle course, or very large in the lower course).

55. Erosion, transportation and deposition

One from:

- traction – where large boulders roll along the river bed
- saltation – where smaller pebbles are bounced along the river bed, picked up and then dropped as the flow of the river changes
- suspension – where finer sand and silt particles are carried along in the river, making it look brown
- solution – where minerals such as calcite from limestone rocks are dissolved and carried along in the flow.

56. Upper course features

In the upper course of a river, near its source, the river has little power to erode rocks. Therefore, it flows around the harder, more resistant rock. This creates spurs that interlock on either side of the valley as the river moves downstream.

57. Lower course features 1

Deposition is the main process in the lower course of a river because the river is carrying a huge load of material, creating landforms such as oxbow lakes, floodplains and levées. There is little downward erosion at this stage as the gradient is very gentle. Deposition and erosion are both important in the formation of meanders.

58. Lower course features 2

D Levée

59. Processes shaping rivers

There are two main types of slope processes. Soil creep is where soil slowly moves down valleys under the influence of gravity **(1)**. Slumping is where the valley sides are eroded by the river, increasing the downward movement of material **(1)**. In both cases, soil, rocks and other material will eventually reach the river, increasing its sediment load. This will impact on river landforms downstream, where the river deposits its load and forms landforms such as meanders, floodplains and levées.

60. Storm hydrographs

Any two from:

- geology – the more resistant the rock, the quicker and greater the run-off, making the rising limb steep and lag time short
- soil type – areas of less soil and more impermeable soils will also increase run-off, making the rising limb steep and lag time short
- slope – the steeper the slope, the faster the run-off, making the rising limb steep and the lag time short
- drainage basin shape – a wide basin with many tributaries close together will make the rising limb steep and lag time short
- antecedent conditions – if the ground is already saturated with water due to previous rainfall, this will affect surface run-off from the subsequent storm, reducing lag time.

61. River flooding

The answer will depend on the located example but factors could include:

- urbanisation increasing run-off from buildings, roads and paved areas
- building on floodplains
- deforestation – removing trees and other vegetation increases run-off
- dredging and channel straightening upstream, increasing flooding risks downstream.

62. Increasing flood risk

Any two from:

- death and injury
- damage to homes, homelessness, etc.
- damage to businesses and loss of revenue
- damage to transport links
- damage to crops and problems with food supply
- damage to freshwater and electricity supplies
- damage to property and potentially to health from pollution carried by floodwater.

63. Managing flood risk

For any suitable soft engineering method, one benefit and one cost should be given. For example:

- River restoration – costs include people living and using the land not wanting it to change, and it can be expensive and difficult to do. Benefits include reducing flooding downstream, and it looks attractive and supports wildlife.
- Washlands – costs include the land cannot be used for anything else, which restricts economic development of the area. Benefits include that it is cheap and requires little maintenance, and it supports wildlife habitats and provides somewhere for the floodwater to go.
- Floodplain retention – costs include the causing of planning problems elsewhere, the restriction of town expansion, and the restriction of economic development. Benefits include the provision of space for recreation and the reduction of properties at risk from flooding.
- Afforestation – costs include the land cannot be used for anything else and it can only be done in some places. Benefits include adding attractiveness to the area, supporting wildlife habitats, and reducing the amount of water reaching the river.

Fieldwork: rivers

64. Investigating rivers: developing enquiry questions

The answer to this question will depend on your own fieldwork location. Being able to compare sites is often a key advantage. If you are comparing the impact of changes in channel characteristics, then you will have aimed to minimise the impact of other factors that might also affect your results.

65. Investigating rivers: techniques and methods

The image of measuring a river depth with a tape measure and ranging pole is a quantitative fieldwork method. It generates numerical data. Collecting people's views by questionnaire is a qualitative method. It generates information that is not numerical but helps with understanding the geographical issue.

66. Investigating rivers: working with data

The main advantage is that it's a good technique to use for comparing data between sites because it shows the range of a set of data and whether the data tend to group together or disperse.

The UK's evolving human landscape

67. Urban and rural UK

Any two from:

- providing tax breaks (Enterprise Zones) and special facilities for businesses to help them grow
- working with the EU's regional development fund to make sure people can access EU funds for businesses (the EU encourages, in particular, innovation in business, business start-ups for women, and businesses that help protect the environment)
- devolving (spreading out) power to local authorities so they have the ability and money to invest in local businesses and infrastructure; improving infrastructure (e.g. rural broadband).

68. The UK and migration

The European Economic Area (EEA) allows citizens free movement between its countries, which means that people from other EEA countries (e.g. Poland, Germany, Ireland or Romania) can come and work in the UK without needing a work permit, as well as enjoying other related benefits. This policy has led to a significant increase in Eastern European populations within some UK places, which increases ethnic diversity, for example. Although many Polish people lived in the UK already, numbers have increased significantly in some areas. This also increases cultural diversity – for example, UK supermarkets now include Polish food sections, introducing different foods and recipes to other UK residents.

69. Economic changes

One reason for the decline in primary sector employment would be that agriculture has become very highly mechanised and computerised. Far fewer people are now needed to farm than in the past.

A second reason for the decline of primary sector employment is the reduction in mining in the UK. When UK-mined coal was the UK's main source of energy, mining employed many thousands of people. As the UK moved to oil and gas, and now towards renewables, the demand for coal reduced. Cheap sources of coal from emerging and developing countries were also used instead of UK coal.

70. Globalisation and investment

It is seen as a disadvantage for countries like the UK because it means that although the TNC may earn millions of pounds of revenue from its UK customers, very little of that revenue comes back to the UK government in tax. This reduces the amount of money the government has for its spending on things like education, the NHS or the military. It also makes it difficult for UK companies to compete with TNCs, because the UK companies pay higher taxes than the TNCs, which reduces profit.

Dynamic UK cities

71. A UK city in context

The answer to this question is specific to your own case study. This will be useful information for writing about change in your city, and identifying land-use characteristics.

72. Urban change differences

The answer to this question is specific to your own case study. If you need to, you can use the Index of Multiple Deprivation explorer to help with this, search for 'IMD explorer' online.

73. City challenges and opportunities

The answer to this question is specific to your own case study. Being able to add relevant details like this about your case study city will help improve your answers.

74. Improving city life

The answer to this question is specific to your own case study. Remember to think about who has been affected by this quality-of-life improvement: is it everyone or just certain groups (e.g. commuters, shoppers, motorists)?

75. The city and rural areas

The answer to this question is specific to your own case study. For example, have commuters moved out of the city to live in the rural area? Has there been a decline in rural services like shops or petrol stations because so many rural inhabitants choose to go to the town rather than buy where they live? Has there been an increase in some rural services geared to commuters, such as nurseries for preschool children, garden centres, horse-riding lessons?

76. Rural challenges and opportunities

Two environmental impacts could include: traffic congestion on the roads around the area as people come from the city to visit the golf course or other attractions of the area; converting farmland to a golf course reduces wildlife habitats and might lower biodiversity in the area. Environmental impacts could be positive too – the map extract includes seven disused pits, suggesting that this might have been an area of quarrying or mining – the golf course could have been landscaping for the area after mining finished here.

Fieldwork: urban

77. Investigating dynamic urban areas: developing enquiry questions

The answer to this question will depend on your own fieldwork location. A key consideration is usually comparability: the ways in which you can minimise factors that could interfere with the results, such as one ward being considerably bigger or smaller than all the others.

78. Investigating dynamic urban areas: techniques and methods

Possible answers would include: burglar alarms, security lights, high fences, remote-controlled gates, metal bars on windows, metal shutters over windows and doors, CCTV on the street, security cameras on homes or businesses, layout of the built environment (e.g. houses facing each other in a semi-circle).

79. Investigating dynamic urban areas: working with data

A combined radar graph would have allowed for a more effective comparison between her five surveyed locations: plotting each one separately was time-consuming and not as easy to use as one combined graph.

Fieldwork: rural

80. Investigating changing rural settlements: developing enquiry questions

The answer to this question will depend on your own fieldwork location. A key consideration is usually comparability: the ways in which you can minimise factors that could interfere with the results, such as one parish being considerably bigger or smaller than all the others.

81. Investigating changing rural settlements: techniques and methods

Possible answers would include: burglar alarms, security lights, high fences, signs on gates warning about dogs, signs warning about DNA traps or dog patrols, remote-controlled gates, security cameras on homes or businesses.

82. Investigating changing rural settlements: working with data

A combined radar graph would have allowed for a more effective comparison between her five surveyed locations: plotting each one separately was time-consuming and not as easy to use as one combined graph.

Extended writing questions

83. Paper 2

There are different ways you could answer this question and there is no one 'right' judgement to come to. Your answer will probably include some of the following points.

For Assessment Objective 3 you should show your geographical understanding of the reasons why some cities grow faster than other cities.

Points you might include for Assessment Objective 3

- People are attracted to cities primarily because of the jobs they offer: that there are more jobs, that the jobs are higher paid than elsewhere and that there is a wider range of opportunities.
- Some cities are particular centres of growth industries – for example, finance or IT. These cities may experience faster growth in jobs, new businesses, and population, as the industries they specialise in also grow.
- Other cities may have been more reliant on secondary sector industries and, as those industries have declined, may not be as attractive to people looking for high paid, rewarding jobs.
- The people who are attracted to the jobs that cities provide may be from other parts of the UK, but also from other parts of Europe and other parts of the world. There are sometimes particular reasons why migrants from one country prefer to settle in a particular UK city: often to do with the existence of a particular kind of work (e.g. agriculture and food processing) and the existence of a community of people also from their country.
- Remember that the question is not just about reasons why cities grow, it is also about the causes of variations in urban growth, so you need to focus your answer on comparisons: what are the factors that have caused some cities to grow more than others, or some cities to decline while others have expanded?

For Assessment Objective 4 you will need to use your geographical skills to provide evidence for your analysis and help you communicate your argument.

Points you might include for Assessment Objective 4

- The diagram shows eight different UK cities and your answer should make direct reference to at least some of them. London is a good city to refer to when talking about economic growth; you may not know many details about Stoke but you can *suggest* reasons why Stoke has not grown as fast as London.
- When you are talking about the reasons why a lot more people might move to a city, then Peterborough would be a good example to use from the diagram. Again, you do not have to know the actual reason for Peterborough's rise in population, you just need to tie it in to your answer to show that you are including data from the diagram in your assessment. For example, you might say, 'Some cities grow rapidly because of a high rate of natural increase, migration from other parts of the UK or immigration from other countries. Peterborough would be a good example of this, with its 15.2% increase in population between 2004 and 2013.'
- Similarly, you could use Newport as an example when you are talking about economic decline of an urban area. You may not know much about Newport, but you can use it to show how you are integrating data from the diagram into your assessment. For example, 'Some cities can experience unemployment from a decline in the number of jobs available. This decline is often associated with urban areas which have relied heavily on secondary sector industries. Newport is an example of a city where the number of jobs has declined by 8.6 per cent between 2004 and 2013.'

COMPONENT 3: MAKING GEOGRAPHICAL DECISIONS

People and the biosphere

84. Distribution of major biomes

1 **D** Tundra

2 Your two differences could include:

- The climate graph for the tropical rainforest (Iquitos) shows that temperature does not vary much throughout the year; it is constantly above 25 °C.
- In contrast, the climate graph for the tundra biome (Kazachye) shows a very high temperature range, with cool summers reaching around 15 °C but extremely cold winters reaching −35 °C.
- The precipitation in the tropical rainforest is high, always above 150 mm in every month.
- In contrast, precipitation in the tundra is low (below 40 mm in the wettest months), and in some months it is very low (below 5 mm).
- Both climate graphs show evidence of a wetter season and a drier season, but no season in the tropical rainforest is ever dry.

85. Local factors

Human activity – a big square has been cleared out of the tropical rainforest by humans.

86. Biosphere resources

Energy resources.

87. Biosphere services

Answers could include:

- regulation of water cycle
- maintenance of soil health
- regulation of atmospheric gases (soaking up carbon dioxide and generating oxygen)
- source of medicines, fuel and building resources.

88. Pressure on resources

Urbanisation As large numbers of people become concentrated in one area, the local water supply comes under a lot of pressure as everyone wants water to drink, wash and clean with. Often cities quickly exhaust locally available water (e.g. from aquifers beneath the city) and have to transport water from other regions, which is expensive and adds to the cost of city living.

Industrialisation Many industrial processes require large volumes of water as part of manufacturing. Even if industries do not need water to make their products or provide their services, industries require a lot of power and power stations require very large amounts of water for cooling purposes.

Forests under threat

89. Tropical rainforest biome

Abiotic: rainwater, topsoil
Biotic: pigs, leaf litter, insects

90. Taiga forest biome

Two negative impacts could include:

- by building roads which attract wildlife, more animals could be killed by collisions with motor vehicles
- roads could interrupt migration routes for some animals
- roads could make it easier to access remote parts of the taiga, allowing loggers to move in and cut taiga forests.

91. Productivity and biodiversity

In the tropical rainforest, trees have shallow roots because nutrients in the soil are quickly lost through leaching; nutrients are concentrated in the topsoil and are taken up very rapidly by biomass. In the taiga, nutrients are concentrated in the litter layer and are released into the topsoil only very slowly; trees have shallow roots in order to access these nutrients as soon as they are released. (Deeper layers of the soil may also be permanently frozen.)

92. Tropical rainforest deforestation

Climate change means increasing global temperatures, which has the impact on equatorial climates of shifting the wet-season Intertropical Convergence Zone, northwards. This reduces the amount of wet season rain reaching rainforests and will make them both hotter and drier. Hotter and drier conditions make forest fires more likely, especially if the soil underneath forests is peat – when peat dries out and then catches fire it can smoulder for months, constantly igniting new fires on the surface.

93. Threats to the taiga

1 reduces the demand for new paper, which reduces the demand for more softwood to be cut in the taiga; **2** reduces carbon dioxide emissions, which helps limit the global warming indirectly threatening the taiga.

94. Protecting the tropical rainforest

Rainforest 'goods' threatened by deforestation include:

- clean water (forests act to filter water and purify it)
- oxygen (rainforests produce a significant proportion of the planet's oxygen)
- future supplies of hardwood timber
- food for those who live in the forest
- drugs and medical treatments (many of which may still be as yet undiscovered).

Rainforest services threatened by deforestation include:

- maintaining biodiversity
- soaking up atmospheric carbon dioxide (buffering climate change)
- protection of soil surface from erosion
- cultural services for tourists
- sense of cultural identity for local people.

95. Sustainable tropical rainforest management

Your answer could include one of the following:
Difficulties in achieving sustainable forest management are likely to include the challenge of preventing illegal logging of hardwood timber, which is very valuable. Local people involved in illegal logging stand to make far more money than sustainable forest management jobs could ever pay – certainly enough also to bribe local police and government officials to allow the logging to continue.

A common difficulty is that sustainable forest management needs to involve local people, but the number of jobs that can be created in this area is always relatively small. That means there will often be a large number of local people who do not benefit directly from sustainable management, which can cause resentment.

Funding can be difficult. Sustainable rainforest management often depends on external funding supplies, such as government money and international aid. This makes it vulnerable to changes in funding decisions.

Population growth is a challenge. If a rainforest region is experiencing population growth, the rainforest will come under increasing threat of deforestation from the expansion of subsistence agriculture and fuelwood collection.

96. Protecting the taiga

Two disadvantages could include:

- It is a very remote area and building the infrastructure to transport the oil to where it is needed is enormously expensive.
- The climate is challenging in eastern Siberia, so getting people to move to the region to work there will also be challenging. A further disadvantage would be the risk of environmental damage to fragile taiga ecosystems: pollution by oil spills, and increased risk of forest fires.

Consuming energy resources

97. Energy impacts

Two negative impacts could include:

- for nuclear – the risk of accidents following natural hazards, human error or terrorist attack
- for biofuels – much recent deforestation of the tropical rainforest in Brazil has been to clear land for planting sugar cane (used for biofuels), and in Indonesia tropical rainforest has been cleared for oil palm plantations (also used as a biofuel, as well as being used in many other consumer products).

98. Access to energy

Three ways could include:

- for wind power, climate could influence how often there was enough wind to generate power from wind turbines
- for solar power, climate could influence how often there were clear skies for solar power generation
- for hydroelectric power, the amount of precipitation through the year would influence the amount of water available for generating hydroelectric power or keeping reservoirs topped up.

99. Global demand for oil

The BP oil spill near Florida reduced oil supply to America, which has a high demand for oil. Reduced supply plus high demand means other supplier countries could raise oil prices. The Arab Spring revolutions were not all in the major oil-producing nations. However, they created instability in the Middle East and, as a result, concern over oil supplies pushed up prices.

100. New developments

One reason could be to increase the UK's energy security: by developing the UK's own gas supplies the UK would not be as dependent on other countries supplying its energy. Alternative reasons could be: fracking could make energy cheaper in the UK; fracking would create jobs in the UK.

101. Energy efficiency and conservation

87 per cent.

102. Alternative energy sources

Energy security is about how secure a country's energy supplies are so that it is able to maintain an affordable price for energy for its citizens. If a country gets all its energy from an unstable region of the world and conflict there affects its supply, it will not be able to maintain affordable energy prices – its energy supply will not be secure.

103. Attitudes to energy

The TNC oil company might have had the view that an exhibition on climate change should present a balanced view that included information on the importance of oil to the economy and why it was important that the UK had access to secure oil supplies for the future. Environmental campaigners may have thought it was wrong for a major oil TNC to influence the Science Museum about what to put in a climate change exhibition: they may have wanted the public to understand their position – for example, that global warming can only be prevented if oil now stays in the ground (a position the oil TNC might not agree with).

Extended writing questions

104. Paper 3 (i)

There are different ways you could answer this question and there is no one 'right' judgement to come to. Your answer will probably include some of the following points.

For Assessment Objective 3 you should show your geographical understanding of the reasons why some groups would be in favour of the development of shale gas.

Points you might include for Assessment Objective 3

- Economic benefits are likely to be top of the list. You could consider the likely groups to benefit: companies that do the fracking (when they sell the gas), people who get jobs in the fracking industry, possibly communities who benefit financially from having fracking developments nearby. The government is also likely to benefit from fracking because it will get taxes from the sale of shale gas.
- The UK government would also benefit from increased energy security if shale gas could meet a large part of the UK's energy needs. This would be because the UK would not then need to import so much energy from other countries, some of which are politically unstable.
- If shale gas was discovered in parts of the country that have problems with unemployment, then fracking could be a source of jobs. This would make it popular with many groups of people even if there were side effects to its development.

For Assessment Objective 4 the job is to use your geographical skills to provide evidence for your analysis and help you communicate your argument.

Points you might include for Assessment Objective 4

- The information about UK government and local government views include interesting points that you can build into your assessment – for example, energy security, economic growth and jobs are three solid areas to explore for **Assessment Objective 4** that you can link directly to the UK government from the table.
- Although the information from fracking companies addresses the environmental impacts of shale gas development rather than reasons why the fracking company

would be in favour of shale gas development, you should make sure you give reasons why the companies would be in favour of the development of shale gas and why it is important to them therefore that they can persuade local communities to allow fracking to take place.

- Because this is an assessment question, you can use the information in the table to assess which group's reasons are the most important.

105. Paper 3 (ii)

There is no single right answer to the options questions: you need to justify why your choice is the one you think has the most going for it. Remember that justify means that you must use evidence to back up your argument, so consider what the best forms of revision effectiveness evidence might be! Percentage of the topic recalled, perhaps?

SKILLS

106. Atlas and map skills

The distribution shows that tropical rainforests are found only in a narrow zone centred on the equator. This suggests that they need warm temperatures and low seasonal variation to develop. This is also an area of the global atmospheric circulation system dominated by the Intertropical Convergence Zone, bringing high rates of precipitation. In order to explain why there is not a continuous belt of tropical rainforest you could talk about local factors – for example, high mountains in South America to explain why tropical rainforests do not extend completely from east to west over the continent. Human activity could be another reason – deforestation – to explain why distribution within this tropical zone is uneven. More than two reasons have been provided here but remember in the exam to only provide the number of reasons that is requested.

107. Types of map and scale

A A lowland river on a wide floodplain.

108. Using and interpreting images

Your two descriptions could include points like: identifying developing tropical cyclones in a source area, monitoring the track of a tropical cyclone over time, monitoring the speed of tropical cyclone movement, estimating likely landfall(s), gauging the damage done by tropical cyclones after landfall.

109. Sketch maps and annotations

Advantages could include that annotated diagrams are effective for showing change over time – for example, in beach profiles or river channel characteristics – which is often difficult to explain in words alone.

110. Physical and human patterns

Settlement Y is located at a crossroads, so access is easy, which would encourage nucleated settlements to develop. It is also located close to another large area of development, such as settlement Z, so settlement Y could be a commuter area. While settlement Z has limited space to expand because of its location in a meander loop, settlement Y does not have restrictions and can expand more easily.

111. Land use and settlement shapes

Use evidence from the map to make your points, which might include:

- there are only isolated pieces of woodland
- there are woods on the south bank of the river to the west of Sellack and next to the court at Baysham
- there is a small plantation of mixed woodland at 565297
- there is another small area of woodland around the road junction at Pennoxstone.

112. Human activity and OS maps

You could include two answers from the following:

Facility	Symbol
Car park	P
Museum	M
Information centre	i
School	Sch
Church	

113. Map symbols and direction

1 public telephone; **2** north-west.

114. Grid references and distances

1 2 km (to the nearest whole number); **2** mixed woodland; **3** 112911.

115. Cross sections and relief

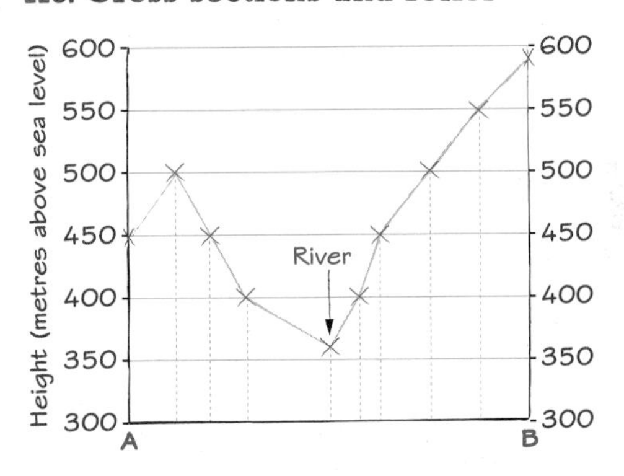

116. Graphical skills 1

(a) line chart; **(b)** scatter graph; **(c)** pie chart.

117. Graphical skills 2

River flow or river discharge.

118. Numbers and statistics 1

Population increase is worked out by:

- finding the increase: new number − original number so 21 million − 8 million = 13 million
- dividing the increase by the original number and × 100 so 13 million ÷ 8 million = 1.625 × 100 = 162.5%

119. Numbers and statistics 2

The interquartile range is the difference between the upper quartile and the lower quartile.
The lower quartile is worked out on the page as being 5 and the upper quartile as 11.
11 – 5 = 6

Notes

Notes

Notes

Published by Pearson Education Limited, 80 Strand, London, WC2R 0RL.

www.pearsonschoolsandfecolleges.co.uk

Copies of official specifications for all Pearson qualifications may be found on the website: qualifications.pearson.com

Typeset and illustrations by Kamae Design, Oxford
Produced by Cambridge Publishing Management Ltd
Cover illustration by Miriam Sturdee

First published 2016

19
10 9 8 7 6 5

British Library Cataloguing in Publication Data
A catalogue record for this book is available from the British Library

ISBN 978 1 292 13378 2

Printed by L.E.G.O. S.p.A.

Acknowledgements

Content is included from David Flint, Kirsty Taylor and Anne-Marie Grant.

The author and publisher would like to thank the following individuals and organisations for permission to reproduce photographs:

(Key: b-bottom; c-centre; l-left; r-right; t-top)

123RF.com: 123rf.com 99tl; **Alamy Images**: Aerial Archives 30c, All Canada Photos 90/2, Andrew Woodley 20, Angela Hampton Picture Library 65br, Bubbles Photo Library 52r, Corporate 74, David Gowans 97bl, 109, DAVID PEARSON 23c, David R Frazier Photolibrary 95, David Speight 39, Dinodia Photos 32, 33t, 34tr, 34bl, eye35 30tr, FLPA 65bl, Frans Lemmens 29t, George S de Blonksy 93br, GraficallyMinded 23r, Ilja Dubovski 90, imagegallery 2 6, Images of Birmingham, Premium 71, incamerastock 47br, Jane Tregelles 59, Jim West 28, Jordan Mary 34cl, Julio Etchart 33b, Marco Palladino 34tl, Marmaduke St. John 99tr, Nick Upton 47bl, Paul Broadbent 97cr, Paul Glendell 63br, Paul Springett 05 87, Photoshot License 93cr, PIB 24r, Rob Hawkins 52l, Robert Harding 97cl, Sabina Jane Blackbird 23l, SCPhotos 29c, Seaphotoart 47tl, Shangara Singh 29r, Sue Cunningham 85, Travelmania 56, Zoonar / GmbH 97br; **Digital Stock**: 88l; **Digital Vision**: 89; **Fotolia.com**: Cloudia Newland 108tr, Ian Woolcock 46tl, Jo Chambers 41, miket 108tl, numage 46tr; **Getty Images**: Digital Globe 7l, 7r, Dream Pictures 21, Geography Photos 48, Keren Su 25, Maximillian Stock 88r, Nigel Roddis 69, The India Today Group 24l, Tobias Schwarz 24c, Yoshikazu Tsuno 102; **NASA**: NASA Earth Observatory image by Jesse Allen and Robert Simmon using EO-1 ALI data courtesy of the NASA EO-1 team. 86; **Rex Shutterstock**: Alban PIX Ltd 50cl; **Science Photo Library Ltd**: NASA 108cl, NOAA 107; **Shutterstock.com**: Brendan Howard 37, Brisbane 63tc, Elzbieta Sekowska 63bl, George Green 50tl, Igor Koza 50tr, Kzenon 99tc, Nickolay Vinokurov 108cr, Steve Heap 47tr, tiorna 78, Wollertz 108br; **UNAVCO**: 15

All other images © Pearson Education

Picture Research by Alison Prior

We are grateful to the following for permission to reproduce copyright material:

Figures
Figures on page 17 adapted from The Congo population pyramid and The UK population pyramid, CIA World Factbook; Figure on page 22 adapted from http://blog.euromonitor.com/2014/06/the-patterns-of-world-trade.html, © 2014 Euromonitor, by permission of the author Virgilijus Narusevicius and Euromonitor International; Figures on page 26 adapted from Towards a more urban world (GMT 2), http://www.eea.europa.eu/soer-2015/global/urban-world, European Environment Agency 2015; Figure on page 40 from Strathayre Forest supplied by Getmapping; Figure on page 45 adapted from Barcelona Field Studies Centre, www.geographyfieldwork.com, Source adapted from Barcelona Field Studies Centre www.geographyfieldwork.com; Figure on page 67 adapted from Usual resident population of urban and rural areas by age, 2011 Census Analysis – Comparing Rural and Urban Areas of England and Wales. Office for National Statistics licensed under the Open Government Licence v.3.0; Figure on page 68 adapted from Long-term International Migration; Office for National Statistics licensed under the Open Government Licence v.3.0; Figure on page 68 adapted from Main comparisons: Population and Migration, Office for National Statistics licensed under the Open Government Licence v.3.0; Figure on page 68 adapted from Immigration and age structure, CIA World Factbook; Figure on page 83 adapted from http://www.centreforcities.org/reader/cities-outlook-2015/3-city-monitor-the-latest-data/ Centre for Cities. Centre for Cities is a UK-based urban economics think tank. This extract is from its report, 'Cities Outlook 2016'; Figure on page 88 from Our World in Data, Esteban Ortiz-Ospina and Max Roser (2016) – 'World Population Growth'. Published online at OurWorldInData.org. Retrieved from: https://ourworldindata.org/world-population-growth/ [Online Resource]; Figure on page 98 adapted from Electricity generated by HEP. Original data from EIA – International Energy Statistics; Figure on page 100 adapted from US net electricity by energy source International Energy Statistics; Figure on page 100 from Public opposition to fracking grows by William Jordan, https://yougov.co.uk/news/2015/05/19/opposition-fracking-britain-grows/, YouGov UK; Figure on page 101 adapted from Energy Consumption in the UK (2015), Department of Energy & Climate Change, from Energy Consumption in the; UK (2015) Department of Energy & Climate Change © Crown copyright 2015. Contains public sector information licensed under the Open Government Licence v3.0; Figure on page 117 from The UK population pyramid, CIA World Factbook

Maps
Maps on page 4 adapted from Tropical cyclone facts, © Crown copyright 2011, Met Office Contains public sector information licensed under the Open Government Licence v3.0; Map on page 20 adapted from UNCTAD's Bilateral FDI Statistics United Nations, © 2014 United Nations. Reprinted with the permission of the United Nations; Map on page 31 adapted from http://shekhar.cc/category/projects/ http://www.coolgeography.co.uk/A-level/AQA/Year%2013/World%20Cities/Mumbai/Mumbai.htm, © Copyright 1996–2016 Shekhar Krishnan; Ordnance Survey Maps, pp.39, 40, 46, 51, 58, 64, 73, 76, 107, 110, 111, 112, 113, 114 © Crown copyright 2016, OS 100030901 and supplied by courtesy of Maps International; Map on page 64 adapted from Environment Agency's flood map http://maps.environment-agency.gov.uk/, Contains public sector information licensed under the Open Government Licence v3.0. © Environment Agency copyright and database rights 2016. © Ordnance Survey Crown copyright. All rights reserved. Environment Agency, 100026380.Contains Royal Mail data © Royal Mail copyright and database right 2016; Map on page 72 adapted from Population and Migration Topic Report October 2013 Planning & Growth Strategy, Planning & Regeneration www.birmingham.gov.uk/census; Maps on pages 76, 77, 80 and 117 from Indices of Multiple Deprivation explorer: http://dclgapps.communities.gov.uk/imd/idmap.html © 2010 NAVTEQ, © 2010 Intermap, © 2016 Microsoft Corporation, Department For Communities and Local Government Contains public sector information licensed under the Open Government Licence v3.0.

Text
Extract on page 6 adapted from Tropical cyclone facts © Crown copyright 2011, Met Office Contains public sector information licensed under the Open Government Licence v3.0; Extract on page 29 adapted from Employment Situation in Mumbai: An analysis http://www.global-labour-university.org/fileadmin/GLU_conference_2010/papers/44._Employment_situation_in_Mumbai_An_analysis.pdf, D. P. Singh, Professor and Chairperson, Department of Research Methodology, Tata Institute of Social Sciences, Mumbai, India, Dr. D. P. Singh with permission; Extract on page 100 adapted from Poll from YouGov (yougov.com).

Notes from the publisher

1. In order to ensure that this resource offers high-quality support for the associated Pearson qualification, it has been through a review process by the awarding body. This process confirms that this resource fully covers the teaching and learning content of the specification or part of a specification at which it is aimed. It also confirms that it demonstrates an appropriate balance between the development of subject skills, knowledge and understanding, in addition to preparation for assessment.

Endorsement does not cover any guidance on assessment activities or processes (e.g. practice questions or advice on how to answer assessment questions), included in the resource nor does it prescribe any particular approach to the teaching or delivery of a related course.

While the publishers have made every attempt to ensure that advice on the qualification and its assessment is accurate, the official specification and associated assessment guidance materials are the only authoritative source of information and should always be referred to for definitive guidance.

Pearson examiners have not contributed to any sections in this resource relevant to examination papers for which they have responsibility.

Examiners will not use endorsed resources as a source of material for any assessment set by Pearson.

Endorsement of a resource does not mean that the resource is required to achieve this Pearson qualification, nor does it mean that it is the only suitable material available to support the qualification, and any resource lists produced by the awarding body shall include this and other appropriate resources.

2. Pearson has robust editorial processes, including answer and fact checks, to ensure the accuracy of the content in this publication, and every effort is made to ensure this publication is free of errors. We are, however, only human, and occasionally errors do occur. Pearson is not liable for any misunderstandings that arise as a result of errors in this publication, but it is our priority to ensure that the content is accurate. If you spot an error, please do contact us at resourcescorrections@pearson.com so we can make sure it is corrected.